Enrique Meza García

Extrusion of Magnesium-Zinc based alloys

Enrique Meza García

Extrusion of Magnesium-Zinc based alloys

Influence of alloying elements on the microstructure and mechanical properties of extruded Mg-Zn based alloys

Südwestdeutscher Verlag für Hochschulschriften

Imprint
Any brand names and product names mentioned in this book are subject to trademark, brand or patent protection and are trademarks or registered trademarks of their respective holders. The use of brand names, product names, common names, trade names, product descriptions etc. even without a particular marking in this work is in no way to be construed to mean that such names may be regarded as unrestricted in respect of trademark and brand protection legislation and could thus be used by anyone.

Publisher:
Südwestdeutscher Verlag für Hochschulschriften
is a trademark of
Dodo Books Indian Ocean Ltd., member of the OmniScriptum S.R.L Publishing group
str. A.Russo 15, of. 61, Chisinau-2068, Republic of Moldova Europe
Printed at: see last page
ISBN: 978-3-8381-2478-0

Zugl. / Approved by: Berlin, TU, Diss., 2010

Copyright © Enrique Meza García
Copyright © 2011 Dodo Books Indian Ocean Ltd., member of the OmniScriptum S.R.L Publishing group

To my beloved parents

and in memory of my grandfather

Abstract

Enrique Meza García

Influence of alloying elements on the microstructure and mechanical properties of extruded Mg-Zn based alloys

In the present work, the influence of the additions of Zinc, Zirconium and Cerium-Mischmetall on the casting, indirect extrusion processing, microstructural development and resulting mechanical properties of magnesium and magnesium alloys were investigated.

It was found that grain size of the cast alloys is controlled by a grain growth factor (Q) mechanism. The importance of Q factor is related to the possibility of predicting the grain size of the cast billet, a parameter which influenced strongly the deformation response of the alloy during extrusion. Grain size and chemical composition were the principal variables of the billets used for indirect extrusion experiments. Their influence on processing was analysed in terms of the extrusion forces and temperatures measured locally during process. The initial grain size of the cast and homogenised billets showed a linear relationship with the profile temperature. Moreover, a noteworthy relationship was found between steady state forces and profile temperatures, showing a linear relationship specific for a constant extrusion speed. Correlations between *initial grain size-temperature* and *temperature-steady state force* allow us to predict the resulting extrusion temperature and load as function of the initial grain size for a specific alloy.

In order to find a correlation between the deformation response and the resulting microstructure of the extruded profiles, Zener-Hollomon parameter (Z) was determined using process variables. It was found that deformation response is alloy composition dependent and it is correlated with the activation of dynamic recrystallisation. The most important contribution of the determination of Z values is the possibility to correlate alloy dependent deformation variables during extrusion with the resulting microstructure; i.e. the resulting average recrystallised grain size.

Recrystallisation process was analysed by texture measurements on the extruded profiles. They revealed a singular development of texture components in function of alloy composition and Z parameter. This development was correlated with the homogeneity of the recrystallised microstructure and the role of intermetallic particles was discussed. Annealing treatment of selected extruded profiles allowed a subsequent analysis of the static recrystallisation process of these alloys. Mechanical properties of the extruded profiles were correlated with the resulting microstructure. Hall-Petch plots were used to separate the effects of grain size from that of the alloying elements on the yield strength and yield asymmetry. Since yield asymmetry behaviour is also influenced by texture, it was possible to integrate texture information in the plots to understand better its influence on the yield behaviour.

Obtained results in this work show correlations between initial alloy conditions and process variables indicating that the resulting microstructure and mechanical properties can be estimated by an appropriate set-up of selected alloy compositions and process parameters. Most of these correlations were based on proved phenomenological assumptions, which make them reliable values of reference for additional research and development of appropriate magnesium alloys for such processing.

Abstract

Enrique Meza García

Einfluss der Legierungselemente auf die Mikrostruktur und mechanischen Eigenschaften von stranggepressten zinkhaltigen Magnesium-Legierungen

In der vorliegenden Arbeit wurde der Einfluss des Zusatzes von Zink, Zirkon und Cer-Mischmetall auf die Mikrostruktur des gegossenen Vormaterials, auf das indirekte Strangpressverfahren, die Mikrostrukturentwicklung und die mechanischen Eigenschaften von Magnesium und Magnesiumlegierungen untersucht.

Es wurde gefunden, dass die Korngröße der Gusslegierungen durch den Korngrößenwachstumsfaktor (Q) geregelt wird. Die Bedeutung des Q-Faktors steht in Zusammenhang mit der Möglichkeit, die resultierende Korngröße in den gegossenen Billets vorherzusagen, deren Wert stark das Verformungsverhalten der Legierung während des Strangpressens beeinflusst. Die Korngröße und die chemische Zusammensetzung waren die wichtigsten Kennwerten der Billets, die für die Strangpressversuche genutzt wurden. Ihr Einfluss auf den Prozess wurde anhand der Strangpresskräfte und -temperaturen analysiert, die lokal während des Prozesses gemessen wurden. Die Eingangskorngröße der gegossenen und homogenisierten Billets ging linear in den Verlauf der Profiltemperatur ein. Darüber hinaus war ein merklicher Zusammenhang zwischen den stationären Strangpresskräften und den Profiltemperaturen festzustellen, welcher spezifisch für eine konstante Strangpressgeschwindigkeit einen linearen Verlauf zeigte. Korrelationen zwischen Eingangskorngröße und Temperatur sowie Temperatur und stationären Strangpresskräften erlauben uns, die daraus resultierende Strangpresstemperatur und -kraft als Funktion der Eingangskorngröße für eine bestimmte Legierung vorherzusagen.

Um eine Korrelation zwischen dem Umformverhalten und der daraus resultierenden Mikrostruktur der stranggepressten Profile zu finden, wurde der Zener-Hollomon Parameter (Z) unter Berücksichtigung der Prozesskennwerte genutzt. Es wurde festgestellt, dass das Umformverhalten legierungselementabhängig ist und es mit der Aktivierung der dynamischen Rekristallisation korreliert. Die wichtigste Vorgehensweise bei der Ermittlung des Z-Parameters besteht in der Nutzung der Möglichkeit, die legierungselementabhängigen Umformvariablen des Strangpressprozesses mit der resultierenden Mikrostruktur, d.h. der durchschnittlichen rekristallisierten Korngröße zu korrelieren.

Der Rekristallisationsvorgang wurde durch Texturmessungen in den stranggepressten Profilen untersucht. Sie zeigten eine Entwicklung der Texturkomponenten in Abhängigkeit von Zusammensetzung und Z-Parameter auf. Diese Entwicklung wurde mit der Homogenität des rekristallisierten Gefüges korreliert und die Rolle der intermetallischen Phasen diskutiert. Eine Wärmebehandlung der ausgewählten stranggepressten Profile erlaubte eine nachfolgende Untersuchung des statischen Rekristallisationsvorgangs dieser Legierungen. Die mechanischen Eigenschaften der stranggepressten Profile wurden mit dem resultierenden Gefüge korreliert. Hall-Petch Schaubilder wurden verwendet um den Einfluss der Korngröße der Legierungselemente auf die Streckgrenze von der Zug-Druck-Asymmetrie zu separieren. Da das Zug-Druck-Asymmetrieverhalten ebenfalls durch Textur beeinflusst wird, war es möglich, die ermittelte Texturinformation in die Schaubilder zu integrieren, um ihren Einfluss auf das Fließverhalten besser zu verstehen.

Die erzielten Ergebnisse in dieser Arbeit zeigen Korrelationen zwischen Anfangslegierungszuständen und Prozessvariablen, was darauf hinweist, dass sowohl die resultierende Mikrostruktur als auch die mechanischen Eigenschaften durch eine passende Einstellung der ausgewählten Legierungen und der Prozessparameter maßgeschneidert werden können. Die meisten dieser Korrelationen basierten auf experimentell verifizierten phänomenologischen Annahmen, die zuverlässige Kennwerte für weiterführende Untersuchungen und Entwicklungen zur gezielten Eigenschaftseinstellung von Magnesiumlegierungen für den Strangpressprozess liefern.

Acknowledgements

Thanks first to my academic supervisor, Prof. Dr. -Ing. Karl U. Kainer, the head of the Magnesium Innovation Centre MagIC and my group leader Dr. rer. nat. Dietmar Letzig, the head of the Department of Wrought Magnesium Alloys for giving me the opportunity to work at the institute, for their valuable guidances, expertise and supports in bringing this work to its completion. Special thanks go to Prof. Dr. Walter Reimers, the head of the Department of Metallic Materials of the Technical University of Berlin for being my second academic supervisor, and for his valuable suggestions to the dissertation. Special thanks also go to Prof. Dr. -Ing. Claudia Fleck for being the chair of my thesis committee.

Special thanks to Dr. rer. nat. Jan Bohlen for introducing me in the magnesium world, his countless suggestions and motivating guidances since the begin of this work. Likewise, I am respectfully indebted to Dr. -Ing. Sangbong Yi for his significant help with the texture determinations, his expertise suggestions and inspiring discussions.

My deep gratitude to Dr. Peter Beaven for his valuable help during the writing and the corrections of this manuscript and his valuable scientific suggestions, which contributed greatly to accomplish this work.

Many thanks to Mr. Volker Kree and Mr. Gerd Wiese for their support in the metallographic preparation and microscope observation. I wish to thank to Mr. Gunter Meister and Mr. Alexander Reichert for assisting me in the foundry; my (actual and former) office colleagues, Rodolfo, Marcus, Ligia, Kerstin, Gerrit, Achim, Lenka, Jacek and Lennart, for such a nice and friendly atmosphere.

The financial support from the Mexican Council of Science and Technology (CONACyT) and the German Academic Exchange Service (DAAD) are acknowledged for providing me the opportunity to do this project.

Finally, I wish to thank to my parents and siblings who were always beside me and Hoa for her love and support.

Table of contents

	Page
1. Introduction...	1
1.1. Aims of the work...	3
2. Fundamentals..	5
2.1 Magnesium...	5
2.1.1 Magnesium as structural material.............................	5
2.2 Deformation behaviour of magnesium..........................	6
2.2.1 Twinning mechanisms during the deformation of magnesium......	9
2.3 Influence of alloying element additions on magnesium....	11
2.3.1 Zinc..	11
2.3.2 Zirconium..	13
2.3.3 Rare Earths (Cerium-Mischmetall).........................	15
2.4 Extrusion of magnesium alloys.....................................	18
2.4.1 Influence of Zn on the extrudability of magnesium..................	19
2.4.2 Influence of Zr and RE on the extrudability of Mg-Zn alloys.......	20
2.4.3 Influence of the alloying elements on the mechanical properties of extruded profiles...	21
2.5 Recrystallisation in magnesium and it alloys.............................	23
2.5.1 Dynamic recrystallisation mechanisms...............................	24
2.6 Development of texture in extruded magnesium alloys...............	27
2.6.1 Contribution of dynamic recrystallisation to the texture development...	31
2.6.2 Contribution of static recrystallisation to texture development......	32
2.7 Yield asymmetry in extruded magnesium alloys.........................	33
2.7.1 Yield asymmetry in terms of the relationship grain size – twinning activation...	33
2.8 Experimental fundamentals..	36
2.8.1 Extrusion methods and process parameters..........................	36
2.8.2. Pole figure goniometer and texture analysis	38
3. Experimental Details...	41
3.1 Investigated alloys...	41
3.1.1 Castings..	43

Table of contents

3.1.2 Characterisation of the cast billets	43
3.2 Process	44
3.2.1 Indirect extrusion	44
3.3 Annealing treatment after extrusion	47
3.4 Characterisation of the extruded profiles	47
3.5 Metallographic procedures	49
3.6 Microstructural characterisation	50
3.7 Mechanical testing	51
3.8 Texture measurements	52
4 Results	**53**
4.1 Characterisation of the homogenised billets	53
4.2 Extrusion experiments	60
4.2.1 Extrusion diagrams	60
4.2.2 Profile temperatures and steady state forces	60
4.3 Microstructures of the extruded profiles	64
4.3.1 Pure magnesium and the Z-series alloys	64
4.3.2 ZK-series alloys	68
4.3.3 ZEK-series alloys	73
4.4 Textures of the extruded Mg-Zn based alloys	77
4.4.1 Pure magnesium and the Z-series alloys	77
4.4.2 ZK-series alloys	80
4.4.3 ZEK-series alloys	82
4.5 Mechanical properties of the extruded alloys	85
4.5.1 Pure magnesium and the Z-series alloys	85
4.5.2 ZK-series alloys	89
4.5.3 ZEK-series alloys	92
4.6 Annealing treatments	95
4.6.1 Indirectly extruded ZK-series alloys after annealing for 1 hour at 400°C	95
4.6.2 Indirectly extruded ZEK-series alloys after annealing for 1 hour at 400°C	101
5. Discussion	**106**
5.1 Cast alloys and their influence on the extrusion process	106
5.1.1 Effect of the alloying element additions on the grain size of the	

Table of contents

billets..	106
5.1.2 Effect of alloy composition and microstructure on extrusion processing	111
5.2 Correlation between extrusion process variables and the microstructure of the extruded profiles	115
5.2.1 Deformation response of the studied alloys during indirect extrusion	115
5.2.2 Relationship between process variables and the resulting microstructure	120
5.3 Recrystallisation as result of the extrusion processing (DRX)............	124
5.3.1 Recrystallisation as result of the annealing treatment (Static RX)	131
5.4 Yield strength, yield strength asymmetry and the Hall-Petch relationship...	135
6. Conclusions..	143
7. Bibliography...	146

Table of contents

List of abbreviations

(Very common abbreviations and units are not listed)

$\dot{\varepsilon}_t$	Strain rate during extrusion
θ	Bragg angle
υ	Extrusion ram speed
φ	Extrusion ratio
ε	Strain rate
λ	Wave length
$\Delta\sigma$	Yield asymmetry
σ_0	Friction stress
$\sigma_{0.2}$	Yield strength
σ_F	Mean flow stress
ΔI	Gradient of intensity
A	Elongation
CRSS	Critical resolved shear stress
CYS	Compressive yield strength
d	Grain size
D_b	Billet diameter
D_e	Profile diameter
DRX	Dynamic recrystallisation
F	Extrusion force
F_{SS}	Steady state force
k	Locking parameter
m.r.d.	Multiples of random distribution
ODF	Orientation distribution function
Q	Growth restriction factor
Q_A	Activation energy for deformation
RDRX	Rotational dynamic recrystallisation
TYS	Tensile yield strength
Z	Zener-Hollomon parameter

1. Introduction

Magnesium alloys have attracted attention in recent years as the lightest available metallic constructional materials. Due to its low density (1.74 g/cm^3) and relatively high specific strength, magnesium offers the possibility to save structural weight, replacing steel and aluminium as construction parts in the transportation industry. The ecological objective of the weight reduction is a decrease in fuel consumption and consequently lower production of CO_2 emissions. Although magnesium and its alloys offer a remarkable potential in this context [1], most current applications involve only cast components i.e. as produced by high pressure die-casting [2]. A much higher potential, which is not readily used today, is the application of semi-finished components of magnesium alloys, such as extruded profiles, sheets or forgings. They show more homogeneous microstructures and improved strength and ductility compared to those of cast components. Nevertheless, technical and economic factors inhibit widespread application of semi-finished components.

Concerning the extruded profiles, magnesium alloys are just being introduced commercially for this purpose. Large scale application of magnesium alloys is hindered by a lack of information on the limits of extrusion deformation (process window) of a still reduced number of available wrought alloys. The main magnesium alloy systems available include alloys that contain 2 to 10 % aluminium, combined with minor additions of zinc and/or manganese, e.g. the AZ and AM families. Because the formability of magnesium at low temperatures is restricted as result of its hexagonal close packed (hcp) structure, it is preferred to carry out extrusion of these alloys at temperatures above 200 °C to avoid cracking. On the other hand, if the temperature is increased to near the solidus temperature, the occurrence of hot-cracking will be promoted. Extrusion at low speeds can prevent heating during processing; however this option is definitely rejected because it reduces considerably the chances of commercial competitiveness. As a consequence, the extrusion speed determines the economic efficiency of commercial extrusion products. This factor can be improved by means of alternative processing methods (e.g. hydrostatic extrusion) or by the development of new wrought magnesium alloys, since the permissible extrusion speeds are alloy element dependent.

The other determining factor for the application of extruded profiles is the resulting microstructure, in terms of grain size and texture, since these parameters influence directly the mechanical properties of the extruded profiles. Grain refinement achieved as a result of processing leads to an increase in yield strength. However extruded profiles also show a strong yield asymmetry during tensile and compressive loading, which is undesirable in structural applications. This effect is a consequence of the resulting texture and the restricted deformation modes at room temperature, which make the accommodation of arbitrary plastic deformation difficult. As a consequence, twinning becomes an active deformation mechanism. Twinning, which is the mechanism responsible for the yield asymmetry, is enhanced if the microstructure has coarser grains. Grain refinement can therefore counteract (or even eliminate) the yield asymmetry [3]. Therefore, grain-refined microstructures as a result of processing are an important profile requirement. Additions of zirconium to magnesium alloys have been used to accomplish this task in cast materials; however, in alloys containing other important elements such as Al or Mn the formation of intermetallic precipitates containing zirconium inhibits this grain refinement effect [4].

The texture resulting from the extrusion process is more complex, since it is influenced by the activation of certain deformation modes at elevated temperatures and by the concurrent process of dynamic recrystallisation (DRX). Unfortunately, information on the deformation behaviour and mechanisms of DRX is scarce [5, 6]. These mechanisms are not very well understood and current research is focussed on such topics. Nevertheless, the possibility to control these mechanisms thermally (via process parameters) and chemically (via alloying elements) exists.

In general, additions of alloying elements could lead to two beneficial effects on the extruded profiles. The first concerns the possibility of easier activation of non-basal slip systems, which should improve deformation at room temperature or even lead to isotropic deformation [7]. The second is the possibility of randomisation of the resulting texture. This effect has been shown to enhance the formability [8].

Examples of the beneficial effects obtained by alloying elements in extruded products have been demonstrated for a group of magnesium-zinc alloys. Zn (the most frequently used alloying element after Al) improves the strength and ductility of magnesium at room temperature [9]. These alloys have been very well adapted to the extrusion process, showing improved mechanical properties in comparison to Al containing alloys. This group of alloys includes modifications with other alloying elements such as the grain refiner zirconium and/or rare earths. Rare earths are generally added to improve the high temperature deformation

1. Introduction

behaviour. Moreover, rare earth additions have also been shown to produce randomisation of the resulting textures [7, 8].

Although most of these results have been obtained on experimental alloys, there are some commercial alloys in this group, e.g.: the zirconium containing ZK60 and the rare earth containing ZE10 alloy [10]. A common characteristic of many zirconium containing Mg-Zn based alloys is their inhomogeneous microstructure, a feature which could also lead to inhomogeneous plastic deformation. Microstructural development in these alloys is strongly dependent on the occurrence of recrystallisation during the extrusion process.

1.1 Aims of this work

In order to improve the extrusion processing of such Mg-Zn alloys, it is necessary to establish and understand the influence of the principal alloying elements on the extrusion process, basically on the recrystallisation process developed during extrusion, as well as on the resulting microstructure, texture and mechanical properties of the extruded profiles.

The present work is therefore focused on the influence of the additions of the principal alloying elements i.e. zinc, zirconium and rare earths (Ce-Mischmetall) on the microstructure, texture and mechanical properties of indirectly extruded profiles. The initial effect of the addition of these alloying elements on the microstructure of the cast alloys is also investigated. Commercially pure magnesium has also been included in this work in order to separate specifically the effects of the alloying elements.

The main topics where the influence of the alloying elements is to be investigated are:

a) On the microstructure of the cast alloys manufactured previous to extrusion process.
b) On the thermo-mechanical response of the alloys during the extrusion process. This includes a correlation with the development of extrusion temperature and load, as locally measured process parameters.
c) On the resulting microstructure of the extruded profiles in terms of the grain size, microstructural homogeneity and texture.
d) On the resulting mechanical properties, basically on the yield strength and yield asymmetry as a function of the resulting texture after the extrusion process.

1. Introduction

e) On the texture evolution in terms of the dynamic recrystallisation process taking place during extrusion.

f) On the texture evolution in terms of the static recrystallisation mechanism taking place during annealing heat treatments on the extruded profiles.

In the topic a) the effect of the alloying element additions on the microstructure, basically their grain refinement effect during casting will be analysed. This will be discussed in terms of inoculation mechanisms by mean of the Growth Restriction Factor Q.

The separate effect of the initial grain size on the extrusion process is also considered. The topics b) and c) will define exactly the role of the initial feedstock conditions in terms of microstructure and composition during processing, specifically on the extrusion load and temperature measured locally during the process. The extrusion temperature is an important parameter for the development of dynamic recrystallisation and subsequent grain coarsening. The understanding of how the process temperature and alloying elements affect these mechanisms is the key to improving the extrusion process for these alloys. Texture measurements on the extruded profiles will show clearly the influence of the alloying elements on the recrystallisation process.

The topic d) will define the influence of the alloying elements and the homogeneity, grain size and texture of the extruded profiles on the resulting yield strength and yield asymmetry. It is important to mention that the influence of some of these elements on the mechanical properties of magnesium has been investigated in single crystals [11, 12] and polycrystalline cast materials[13-15], but little information on their influence in textured extruded profiles can be found in the literature.

The purpose of the topics e) and f) is to establish the influence of the principal alloying elements on the dynamic recrystallisation (DRX) occurring during extrusion and the static recrystallisation (SRX) occurring during subsequent annealing of the extruded profiles.

The results of this work can be used to optimise processing parameters for extrusion of Mg-Zn based alloys. Moreover, the variation in composition of the alloys investigated will provide the knowledge essential to setting up guidelines for the development of appropriate magnesium alloys for such processing and thus to extend the use of these alloys.

2. Fundamentals
2.1 Magnesium

With a density of 1.74 [g/cm^3] magnesium is the lightest structural metal. This is particularly noticeable if it is compared with the density of the most frequently used commercial structural metals: steel (7.8 [g/cm^3]) and aluminium (2.7 [g/cm^3]). Other important physical properties of magnesium are shown in Table 2.1. The low density of magnesium, coupled with higher strength-to-weight ratio makes magnesium the material of choice for lightweight components. Moreover, its low melting point and low heat of fusion allow good castability and thus efficient production of cast components. On other hand, the very electronegative potential of magnesium in combination with impurities i.e. dissimilar elements, secondary phases or impurities, makes magnesium alloys very prone to galvanic corrosion.

Table 2.1 Some important physical properties of magnesium.

Crystalline structure	*Hexagonal close packet (hcp)*
Lattice parameter a	*0.32094 nm (±0.01%)*
Lattice parameter c	*0.52107 nm (±0.01%)*
c/a ratio	*1.6236*
Atom radius	*0.159 nm*
Melting point	*648.8 °C*
Latent heat of fusion	*382 [KJ/kg]*
Oxidation standard potential	*-2.4 Volts*

2.1.1 Magnesium as structural material

Pure magnesium cannot be used for structural applications because of its softness (37 Brinell hardness number hardness (HBR) [16]) and mechanical weakness (tensile and compressive yield strength of 21 MPa [16]). In order for magnesium to be used in structural components, it is necessary that magnesium exhibits sufficient strength and ductility under dynamic loading as well as static loading conditions at ambient temperature. The strength can be influenced by a combination of well-known hardening mechanisms, i.e. solid solution hardening, precipitates dispersion hardening, work hardening, and boundary hardening. The ductility of magnesium can be considerably improved by grain refinement. These aims can be accomplished by the addition of certain alloying elements. These additions also influence other properties such as reactivity of the melt, castability, and corrosion performance. By careful selection of alloying elements, alloys for general purpose and for special applications

2. Fundamentals

can be produced. The most commonly used alloying elements are aluminium, zinc, manganese, and zirconium. Rare earths and yttrium are used for elevated temperature applications. More details of the effects of some of these on magnesium can be found in Chapter 2.3. Unfortunately, the maximum levels of these elements are limited by their low solubility in magnesium solid solution, which limits possibilities to causes significant changes in, for example, the crystallographic lattice parameters of magnesium. Thus, the deformation behaviour of magnesium alloys is strongly dependent on its hexagonal close packed (hcp) crystal structure, which is an important factor limiting wider usage of magnesium alloys as structural materials.

2.2   Deformation behaviour of magnesium

The deformation of magnesium is strongly influenced by the inherent anisotropy that results from the low symmetry of the hcp structure. This low symmetry limits the availability of different slip systems which can be activated simultaneously. A slip system is defined by a slip plane together with the slip direction. Independent of the crystal structure, the slip plane is the plane of greatest atomic density and the slip direction is the closest-packed direction within the slip plane. Since the planes of greatest atomic density are also the most widely spaced planes in the crystal structure, the resistance to slip is generally less for these planes than for any others set of planes. Slip occurs when the shear stress on the slip planes in the slip direction reaches a threshold value called the *critical resolved shear stress* (CRSS). This slip mechanism is controlled by dislocation glide.

The crystallographic indices of the principal slip planes of the hcp structure of Mg and some of its available slip systems are shown in Fig. 2.1. In hcp metals, the only plane with a high atomic density is the basal plane (0001). The axes $\langle 11\bar{2}0 \rangle$, also called $\langle a \rangle$ are the close-packed directions. For magnesium at room temperatures, slip occurs on the (0001) plane in the $\langle 11\bar{2}0 \rangle$ directions, because the critical resolved shear stress (CRSS) value for basal $\langle a \rangle$ slip is a hundred-fold lower than that required for activation of the $\{10\bar{1}0\}\langle 11\bar{2}0 \rangle$ prismatic slip and $\{10\bar{1}1\}\langle 11\bar{2}0 \rangle$ pyramidal slip (Type I) systems, see Fig. 2.2 [17-19]. Since there is only one basal plane per unit cell and three $\langle 11\bar{2}0 \rangle$ directions, the hcp structure possesses three slip systems. However, basal slip is only able to offer two independent slip systems. Moreover, deformation along the c-axis cannot be accommodated by any of the three slip systems mentioned above, either singly or in combination with the others, because all of them have the $\langle 11\bar{2}0 \rangle$ slip direction. Thus, the two available slip systems are not enough to satisfy the von

2. Fundamentals

Mises criterion for uniform plastic deformation in polycrystalline materials which requires the operation of at least five independent slip systems [20].

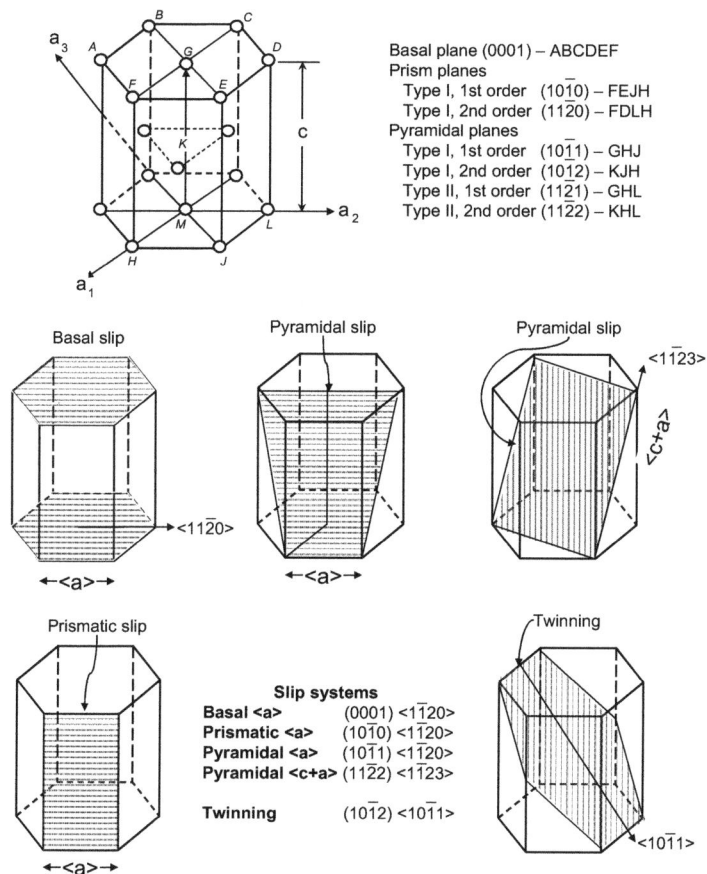

Fig. 2.1. Low index planes and directions in the hcp structure of magnesium. Examples of the most important slip systems with their respective indices are shown [21, 22].

An additional deformation mechanism which can compensate for the lack of slip systems is twinning. However, it represents only a half of an independent slip system [23]. On the other hand, twinning may assist the activation of prismatic and pyramidal slip because it produces a considerable change in the orientation of the crystal, which may place these non-basal slip systems in a favourable orientation with respect to the stress axis. Thus, the activity of

2. Fundamentals

prismatic slip $\{10\overline{1}0\}\langle11\overline{2}0\rangle$ [24] as well as pyramidal slip (Type II) $\{11\overline{2}2\}\langle11\overline{2}3\rangle$ also called $\langle c + a\rangle$ slip system has been reported in studies on deformation at room temperature in magnesium [25, 26]. Moreover, the $\langle c + a\rangle$ slip system is able to accommodate deformation in the c-direction, which should contribute to improve formability.

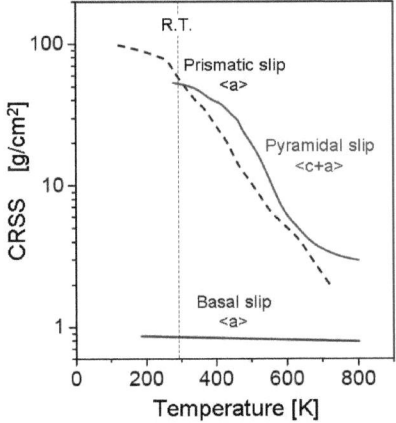

Fig. 2.2. Temperature dependence of the CRSS for various slip systems in pure magnesium [17, 18].

In spite of these possibilities to activate additional slip systems, deformation at room temperature in polycrystalline magnesium is limited only to basal slip and twinning, which results in low formability. If twinning dominates during deformation, it leads to brittle failure [27, 28]. Nevertheless, it is very important to notice in Fig. 2.2, that the CRSS of the slip systems is temperature-dependent [17]. Between room temperature and 200 °C there is a significant and continuous decrease in the CRSS for prismatic slip. At temperatures above 200 °C, the decrease in the CRSS for both prismatic and pyramidal slip is even more pronounced. Thus, enough slip systems become active and the formability of the magnesium improves considerably. This is the reason why the deformation of magnesium alloys during thermo-mechanical processing, e.g. extrusion, is more effective if it is carried out above this temperature. Besides the temperature, additions of certain solute elements e.g. Zn or Li can also contribute to reductions in the CRSS of non-basal slip systems in magnesium [12].

2. Fundamentals

2.2.1 Twinning mechanisms during the deformation of magnesium

Due to the importance of twinning as a deformation mechanism in magnesium, especially at room temperature, it is necessary to describe the most outstanding characteristics of this mechanism. As mentioned above, twinning occurs when slip systems are restricted, this explains the occurrence of twinning at low temperatures or high strain rates in hcp metals at orientations which are unfavourable for basal slip [29].

The lattice strain needed to produce a twin configuration in a crystal is small, so that the amount of total deformation that can be produced by twinning is small [30]. The important role of twinning in plastic deformation does not come from the strain produced by the twinning process but from the orientation changes resulting from twinning. Twinning may place new slip systems in a favourable orientation with respect to the stress axis so that additional slip can take place. Twinning deformation has very specific characteristics that distinguish it from crystallographic slip and they can be summarised as follows:

a) Twinning is a unidirectional deformation mechanism. It can produce either a tensile or a compressive strain along the c-axis, but not both [28].

b) Twinning has a limited capacity for strain accommodation. The amount of plastic shear caused by twinning is small and fixed, only 0.065 in magnesium [30].

c) It is considered that twins can act as barriers to dislocation glide, therefore their massive formation leads to a grain refinement effect [31].

d) Twinning causes a sudden and large reorientation of crystals. The nature of the new orientation depends on the kind of twin system.

In the literature three types of twins have been reported in magnesium alloys: tensile twins, compression twins and double twins. The first two types can be described by Fig. 2.3 where the amount of twinning shear for possible twinning modes is shown as a function of the c/a ratio. If the slope lies on the left side of a shear value of zero, it has a negative gradient, and the twin causes extension along the c-axis. This twin mechanism corresponds to a contraction twin. Conversely, a positive gradient causes contraction along the c-axis, corresponding to a tension twin. If we focus attention on the vertical line representing the c/a ratio of Mg (~1.624), it can be observed that the $\{10\bar{1}1\}$ and $\{11\bar{2}2\}$ twins are compression twins, whereas the $\{10\bar{1}2\}$ twin is activated by tension along the c-axis. Thus the $\{10\bar{1}2\}$ tensile twin is the predominant mode in Mg with the smallest twinning shear and is favourable when

2. Fundamentals

there is an extension strain component parallel to the c-axis or when compression is applied perpendicular to the c-axis. Its twinning shear is 0.1289 and the angle between the basal planes and the twinning $\{10\bar{1}2\}$ planes is 43.15°.

In contrast to $\{10\bar{1}2\}$ tensile twins, $\{10\bar{1}1\}\langle10\bar{1}2\rangle$ compression twins are activated when there is a contraction strain component parallel to the c-axis (also called contraction twins) or when macroscopic extension is applied perpendicular to the c-axis. In this twinning system the basal planes are rotated around the same $\langle11\bar{2}0\rangle$ direction as in $\{10\bar{1}2\}$ twinning, but the rotation angle is 56.2°.

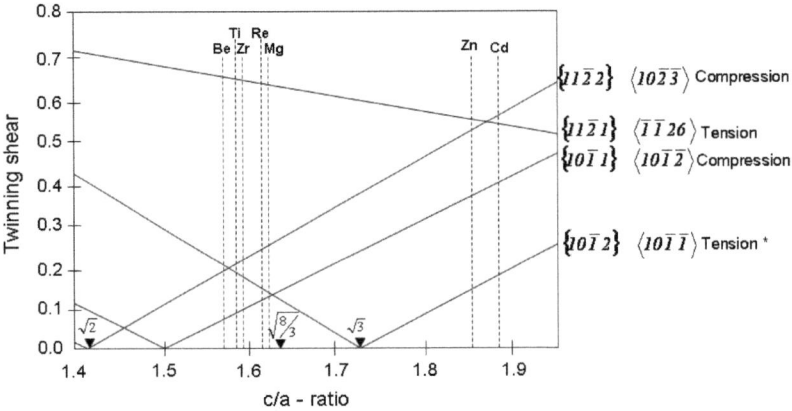

Fig. 2.3. Variation of twinning shear with the axial ratio. For the seven hexagonal metals. * indicates that the twin system is an active mode[28].

Double twinning is a secondary twinning mechanism that can take place within the reoriented primary twins. The first condition to carry out this mechanism is the presence of $\{10\bar{1}1\}\langle10\bar{1}2\rangle$ compression twins, then the material inside the twin undergoes a secondary $\{10\bar{1}2\}$ tension twinning; thus both twinning events produce rotation about the same $\langle11\bar{2}0\rangle$ axis.

The reversibility of twinning is an additional mechanism called detwinning or untwinning. This mechanism occurs in a twinned material and is frequently observed in compression-tension cyclic loading. It is characterised by the disappearance of existing twins bands, e.g. twins are able to disappear or become narrower under reverse loading or unloading and they can even reappear on reloading [32, 33].

2. Fundamentals

2.3 Influence of alloying element additions on magnesium

The most effective way to improve the strength and ductility of magnesium is by alloying with additional elements. The improvement of these and other mechanical properties can be influenced directly by increasing the content of solute element in solid solution, or indirectly, e.g. by a grain refinement effect or by the development of intermetallic phases. Some significant advances in this respect have been achieved with the addition of zinc, zirconium and rare earth elements to magnesium, the alloying elements used in this work. In this section the effects, benefits and limitations of the addition of these alloying elements to magnesium will be described. This information is relevant to the selection of appropriate process parameters for the alloy compositions used in this investigation.

It is important to mention that most of the information compiled in this section refers to cast materials and single crystals. However, information about the effect of these alloying elements on the extrudability and resulting mechanical properties of extruded profiles will be described separately in sections 2.4.

2.3.1 Zinc

Zinc is next to aluminium - one of the most frequently used alloying elements in magnesium. Zn is often used in combination with aluminium to produce improvements in room-temperature strength; however, it increases hot-shortness when added in amounts greater than 1 wt. % to magnesium alloys containing 7 to 10 wt. % aluminium [34]. During melting and casting, Zn improves the fluidity of the melt but may also induce micro-porosity in cast material [35]. Zinc acts as a grain refiner [16, 35, 36] and this may result in an increase of strength as result of the Hall-Petch effect [14, 37]. Zn also helps to overcome the harmful effects of iron and nickel impurities on the corrosion properties of magnesium alloys [34]. The most important ternary additions to binary Mg-Zn alloys are zirconium or rare earths and these will be described in detail later.

As a rule, the zinc contents of these alloys do not exceed the maximum solubility of 6.2 wt. % (2.4 at. % respectively) at 340 °C, see the phase diagram in Fig. 2.4. Since the solubility decreases substantially with temperature - the solubility is 1.1 wt. % (0.5 at. %) at room temperature - decomposition of the supersaturated solid solution of zinc in magnesium can be controlled by heat treatment. This effect can be applied to increase the strength of these alloys by precipitation hardening [38-40].

2. Fundamentals

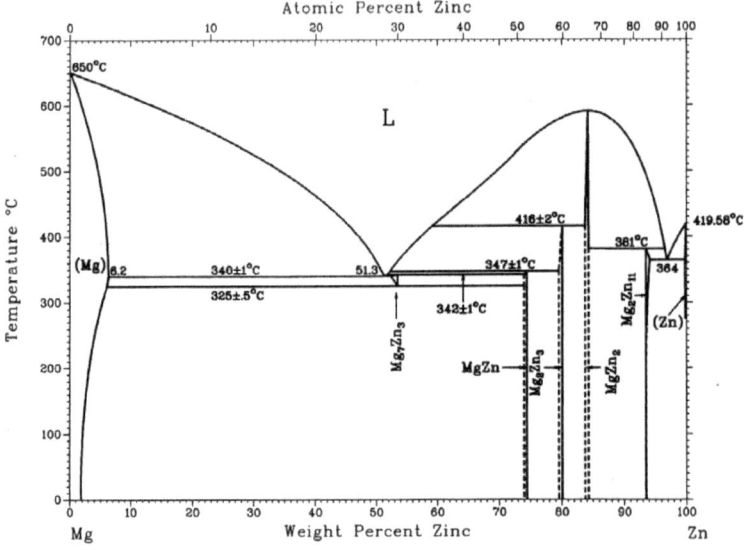

Fig. 2.4. Phase diagram of the Mg-Zn system [41].

On other hand, it has been shown that many kinds of solute atoms, including Zn, produce solid solution strengthening in single crystals as well as polycrystalline cast magnesium materials [11, 12, 14, 42-44]. In single crystals of magnesium with small amounts of solute, solid solution hardening is also anisotropic: solute hardens the basal planes, but softens the prismatic and pyramidal planes.

Akhtar and Teghtsoonian [12] have shown that solution softening of the prismatic planes in dilute alloys is a general effect in magnesium, irrespective of any changes in the c/a ratio or electron density brought about by the solute atoms, and is thus probably connected with a reduction in the stress required to overcome the Peierls-Nabarro lattice friction. The Peierls stress is the shear stress required to move a dislocation through a crystal lattice in a particular direction [22].

In as-cast, polycrystalline magnesium the addition of Zn improves the strength at room temperature. At concentrations up to 1 at.% Zn the ductility increases and this is thought to be connected to the Zn-induced softening of the prismatic planes observed in single crystals [14]. At concentrations above 1 at.% Zn, the strain hardening rate increases and this coincides with

12

2. Fundamentals

a decrease in elongation, suggesting that solid solution softening of secondary slip is gradually offset by solid solution hardening at high solute concentrations [44]. Since Zn and Al result in similar rates of solution hardening of the basal plane in dilute alloys [11, 42], the same behaviour could be expected for polycrystals of concentrated Mg-Zn alloys. However, evidence from the literature suggests that Zn is far more efficient than Al in strengthening Mg at high solute concentrations. In rapidly solidified Mg alloys, it has been reported [14] that Zn hardens Mg twice as much as Al on an atomic % basis. The CRSS's of the basal and prismatic slip systems also exhibit marked differences in their temperature dependence [11, 12]. Therefore the range of temperature between 200 and 550 °C seems to be ideal to take advantage of the solid solution effect on the formability during deformation at high temperatures, i.e. during extrusion.

It is important to mention that effects of Zn additions in textured polycrystalline materials have not yet been investigated. One aim of the current work is therefore to study the role of Zn and texture on the mechanical response at room temperature of extruded profiles of binary Mg-Zn alloys.

2.3.2 Zirconium

Zirconium is a potent grain refiner in magnesium alloys that contain little (impurity level) or no Al, Mn, Si, and Fe (zirconium forms stable compounds with these elements and its grain refinement effect is counterbalanced) [4, 35]. Zinc (and rare earth) containing alloys are therefore the major Mg alloy group grain refined by Zr at the present. It is thought that because the lattice parameters of α-zirconium (a=0.323 nm, c=0.514 nm) are very close to those of magnesium (see Table 2.1), zirconium-rich solid particles produced early in the freezing of the melt may provide sites for the heterogeneous nucleation of magnesium grains during solidification [16, 34].

The maximum solubility of Zr in molten pure magnesium at 654 °C is 0.6 wt. %, see Fig. 2.5. At the peritectic temperature it increases until 3.8 wt. %; still, it drops to 0.3 wt. % at 300°C and remains at this level at room temperature. It has been reported that with Zn additions up to ~ 4 wt. %, the solubility of Zr increases [9, 45], making this alloy system very attractive as it takes advantage of the benefits of both elements. When Zr is added to these alloys (~0.32 wt. %), it can readily reduce the average grain size from a few millimetres to 80 - 100 μm at normal cooling rates [46]. Moreover, well-controlled grain refinement by Zr can lead to the

2. Fundamentals

formation of nearly equiaxed or nodular grains [47], which further enhance the structural uniformity of the final alloy. This exceptional ability of Zr in cast alloys has also been extended to generate even more enhanced grain refinement in extruded profiles [48, 49]. This has led to the development of some commercially important magnesium extrusion alloys, e.g. a high strength extrusion alloy, ZK60 (Mg-6wt. %Zn-0.6wt. %Zr) and a lower zinc content variant ZK40 (Mg-4wt. %Zn-0.6wt. %Zr).

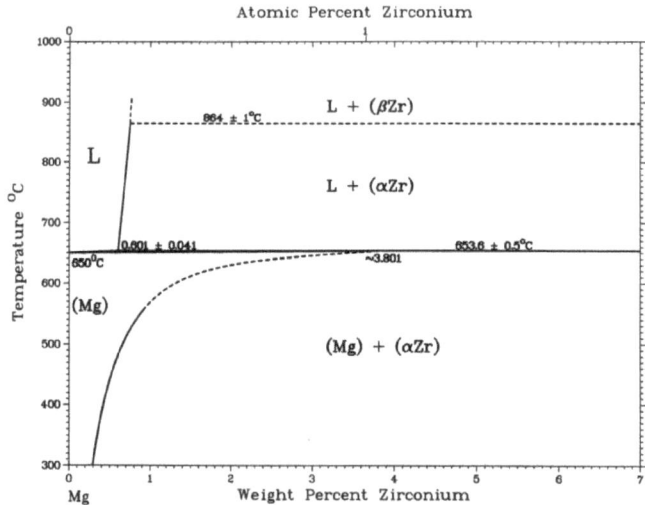

Fig. 2.5. Phase diagram of the Mg-Zr system [50].

The Mg-Zn-Zr system is considered - after the Mg-Al-Zn system – to provide some of the most important magnesium alloys of practical interest. Zr additions to Mg–Zn alloys also increase the solidus temperature, which could be beneficial during hot deformation [9].

Concerning the formation of intermetallic phases, some reports indicate that this system does not develop any ternary intermetallic compounds [51-53]. Most of the intermetallic phases observed in this system have been related to the system Zn-Zr and to a minor degree to the Mg-Zn system. The Zn-Zr phase diagram is shown in Fig. 2.6 (a). This system has the most stable intermetallic compounds. It is very likely that the low solubility of Zn and Zr in the hcp structure of magnesium at low temperatures is the reason for segregation of both solutes. By means of segregation, the solutes combine to form Zn –Zr compounds. The most commonly observed intermetallic phases in the binary Zn-Zr diagram are ZnZr and Zn_2Zr, whereas in an isothermal Mg-Zn-Zr ternary diagram at 300 °C the presence of Zn_2Zr_3 is indicated in Fig. 2.6 (b) [54].

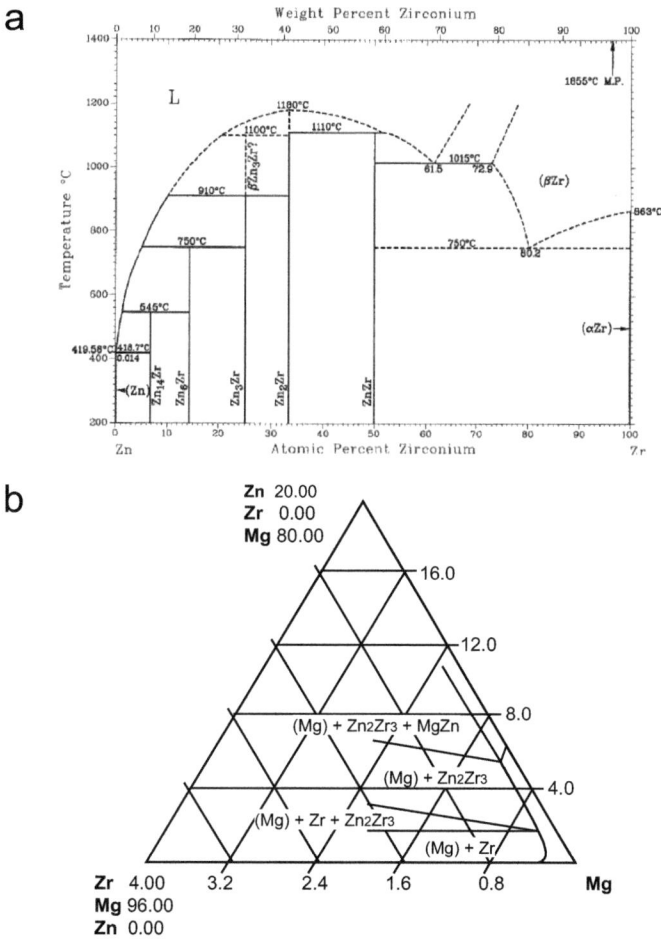

Fig. 2.6. (a) Phase diagram of the Zn-Zr system [50] and (b) Ternary diagram Mg-Zn-Zr (wt. %) isothermal section at 300°C [54].

2.3.3 Rare Earths (Cerium-Mischmetal)

Cerium, lanthanum, praseodymium, neodymium, thorium, yttrium and scandium are among the most widely investigated rare earth additions to magnesium alloys. The application of some of these, for example thorium (which is radioactive), and praseodymium and scandium (which are quite expensive) has been discontinued.

2. Fundamentals

Rare earths have been added to magnesium either individually or in the form of "mixtures". Despite the substantial differences in constitution and properties, magnesium alloys with separate rare earth additions show similar features [55]. All rare earth metals lead to improvements in strength when they are added to magnesium. Moreover, grain refinement and increased ductility are observed even at small contents. These alloys show good casting properties and reduced weld cracking, due mainly to their narrow freezing range (this tends to suppress porosity). The mechanical properties of the alloys depend to a large extent on the possibility of solid solution decomposition in the alloys which results in the formation of a coarse network of complex Mg-Zn-RE intermetallic phases [52, 56, 57]. The good strength properties, in particular, high temperature creep resistance and thermal stability [57], make them attractive for commercial applications.

In spite of these findings, there are only few commercial RE containing wrought alloys, e.g. ZE10 (Mg-1.25wt. %Zn-0.2wt. %Ce-MM), WE43 (Mg-4wt. %Y-3wt. %Nd-0.5wt. %Zr) and WE54 (Mg-5wt. %Y-3.5wt. %Nd-0.5wt. %Zr) [35, 55, 58]. A limitation of RE elements is that they are expensive. To date, the above mentioned "mixtures" of RE metals are generally used to reduce the high costs. The most well-known among them is Cerium – Mischmetal (Ce – MM). Its composition corresponds approximately to the contents of the individual RE metals in the most widespread ore, monazite, and includes mainly cerium, lanthanum, neodymium and praseodymium. An example of a Ce-MM composition is presented in [15] as follows (in % mass): La 22.6, Ce 50.6, Pr 6.4, Nd 18.2, Fe 0.59, Si 0.16, Cr 0.03, and other impurities 1.42. The main RE metal in the composition is cerium with typical contents in the range 50 -75 mass % and the other main element is lanthanum.

One important limitation of the above mentioned alloying elements is their limited solubility in magnesium A short overview of the maximum solubilities and other physical properties of the principal alloying elements used in this work: Zn, Zr, Ce, La, Pr and Nd, is shown in Table 2.2.

2. Fundamentals

Table 2.2: Maximum solubilities of selected alloying elements in magnesium and some of their physical properties.

Element	Crystalline structure	Density [g/cm³]	Atom radius [nm]	c/a	Melting point [°C]	Max solubility in Mg [wt.%]
Zn	hcp	7.14	0.133	1.86	419.6	6.2 at 340 °C
Zr	hcp	6.57	0.159	1.59	1852	0.6 at 654 °C
Ce	fcc	6.77	0.182	-	798	0.74 at 590 °C (1)
La	dhcp	6.146	0.187	1.619	920	0.22 at 600 °C (1)
Pr	hcp	6.475	0.182	1.614	931	1.7 at 575 °C
Nd	hcp	7.004	0.181	1.614	1010	3.6 at 552 °C
Mg	hcp	1.74	0.159	1.623	648.8	-

Abbreviations: hcp, hexagonal close-packed, dhcp, double hexagonal close-packed; fcc, face-centred cubic.
(1) In the form of Ce-MM: the approximate maximum solid solubility in magnesium is between 0.18 and 0.40 [55].

It is important to mention that RE additions to magnesium alloys are of little practical interest unless Zr and/or Zn are added, the first as grain refiner and the second as strengthener. The combined addition of these elements is promising for the development of new, wrought magnesium alloys. However, knowledge on the effects of RE additions during the processing of semi-finished products such as extruded profiles, sheets or forgings is still scarce, although experimental extrusion trials have already been carried out successfully on such alloys [59, 60]. The thermally activated mechanisms of recrystallisation in these alloys have to be understood in order to improve their thermo-mechanical processing.

2.4 Extrusion of magnesium alloys

In this section, the availability of some magnesium alloys adapted to extrusion and the problematic associated with its use, will be described. Information on the influence of the above mentioned alloying elements on processing and resulting magnesium extrusions is included. A description of the extrusion methods and principal process parameters can be found in section 2.8.

Extrusion is one of the basic shape giving technologies for metals and alloys that allows the production of long profiles with uniform and even thin-walled cross sections. Extrusion is the process by which a block of metal is reduced in cross section by forcing it to flow through a die orifice under high pressure. Although the process technology has existed for 60 years [35, 61], only very few applications on magnesium alloys, especially of the Mg-Al-Zn family (called AZ), are available today [62]. Other magnesium alloys that have been commercially applied as extruded products correspond to the systems Mg-Mn (called AM or M), Mg-Zn-Zr (called ZK), Mg-rare earths and Mg-Li [10, 16]. Due to the low formability at room temperature it is preferred to carry out extrusion of these alloys at temperatures above 200 °C, in order to avoid cold cracking and to increase the extrusion speed [35, 58, 63]. Typical industrial process temperatures are in the range from 260 °C to 450 °C [35, 59, 64, 65].

The limits of extrusion for magnesium alloys can be schematically illustrated by the process window diagram shown in Fig. 2.7 [59]. The pressure limit is represented by the locus of a

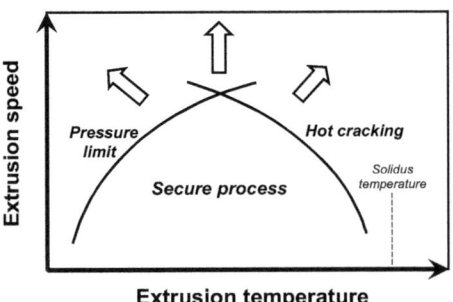

Fig. 2.7. Process window showing the optimum conditions for safe processing of magnesium alloy extrusions [59].

2. Fundamentals

specific extrusion pressure correlated with the press capacity of the machine and the temperature limit is shown as the locus of the incidence of hot cracking. The latter is limited by the solidus temperature in the corresponding phase diagram. The window generated between the two loci shows the safe regime for successful extrusion. The pressure limit can be increased by using a higher press capacity, for example the use of hydrostatic presses [66]. The use of hydrostatic presses has allowed magnesium alloys to be extruded successfully at temperatures between 100 and 200 °C [49, 67, 68].

An important issue in the extrusion of magnesium alloys is the extrusion speed because it will determine the manufacturing efficiency of commercial products. Experiments with magnesium alloys have already shown their capacity to stand high extrusion speeds, e.g. between 40 and 120 m/min during hydrostatic extrusion [48]. The extrusion speed itself is not the principal limitation to the low workability of magnesium, but rather the temperature rise that occurs with increasing speed. This is caused by heat generation from high rate plastic deformation and friction. Due to the negligible heat dissipation from the billet to the extrusion tooling, it is also known as adiabatic heating. This heating can lead to incipient melting and hot-cracking. This is characteristic of the AZ magnesium alloys which have solidus temperatures that decrease significantly with increasing solute content [34]. The presence of second phases with low melting temperature can also increase the propensity to hot-cracking [69].

The adiabatic heating also influences other thermally-activated mechanisms during extrusion, e.g. dynamic recrystallisation and the subsequent grain coarsening. These topics are described in detail in section 2.5.

2.4.1 Influence of Zn on the extrudability of magnesium

For the selection of appropriate process parameters, it is important to take into account the effect of the selected alloying elements on the extrudability of magnesium. Strangely enough, there is scarce information on this topic. One study on the effect of the Zn and Zr contents on the extrusion conditions has been carried out by Doan and Ansel [9]. They found that during direct extrusion, Zn additions greater than 1 wt. % promoted the occurrence of hot-cracking [70]. Fig. 2.8 shows the effect of the Zn addition on the hot-cracking limit in directly extruded Mg-Zn alloys with a constant Zr content. This sharp fall in the hot-cracking limit is due to the decrease of the solidus temperature with addition of Zn solute in magnesium, see Figs. 2.4

and 2.9. An increase in Zn content also promotes the formation of second phases with low melting temperatures and could also contribute to the drastic fall in the hot-cracking limit.

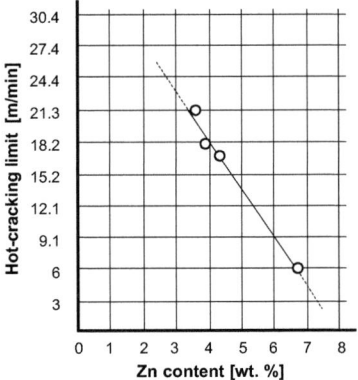

Fig. 2.8. Effect of Zn additions on the extrudability of Mg-Zn alloys [9].

2.4.2 Influence of Zr and RE on the extrudability of Mg-Zn alloys

In the same study, Doan and Ansel reported that Zr additions promoted an increase in the solidus temperature of cast Mg-Zn alloys (see Fig. 2.9), which allowed the use of greater extrusion speeds (Fig. 2.10). However, Zr content in wrought magnesium alloys will not overcome its limit of solubility in molten magnesium (0.6 wt. %).

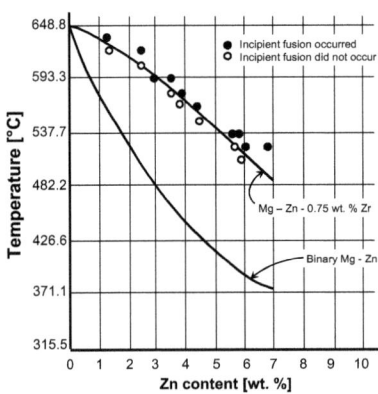

Fig. 2.9. Solidus of Mg-Zn and Mg-0.75wt. % Zr vs. Zn content [9]

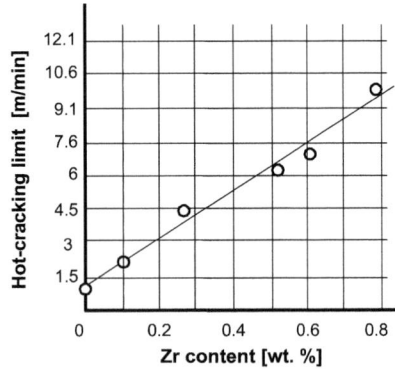

Fig. 2.10. Effect of Zr on the extrudability of Mg- 6 wt. % Zn- Zr [9].

In the case of the rare earths, there are no studies of their effects on the extrudability of magnesium. The successful extrusion trials on RE containing magnesium alloys reported in [56, 71] focused on the microstructures resulting from processing. It was found that ternary or higher order alloying additions may be able to suppress grain growth during extrusion. RE additions resulted in the formation of inter-dendritic precipitates in the billet; these precipitates were broken down into small thermally stable particles during extrusion. Moreover, a homogeneous distribution of fine particles was also induced throughout the matrix during extrusion. Both sets of particles combined to retard grain growth in the extruded product significantly [72].

Cerium, an important addition in the present work, has shown as an individual addition, similar behaviour and several successful trials using direct extrusion at temperatures between 300°C and 500 °C and extrusion ratios between 1:10 and 1:36 have been reported [55]. Higher extrusion temperatures have been also successful used for Zr containing Mg-RE alloys [61]. It is noteworthy that this capacity to extrude RE containing magnesium alloys at high temperatures could also allow an increase in the extrusion speed.

2.4.3 Influence of the alloying elements on the mechanical properties of extruded profiles

Parallel to the scarce bibliographical references about the influence of the alloying elements on the extrudability, there exists also little information about the resulting mechanical properties of the extruded alloys.

In the case of Zn, the work of Doan and Ansel will be referenced again. Their results are summarised in Fig. 2.11, where the effects of the solute content and extrusion speed on the yield asymmetry are shown. The yield asymmetry is represented by $\Delta\sigma = TYS - CYS$, (where CYS is the compressive yield strength and TYS the tensile yield strength) at constant initial grain size, noting that the lowest value of $\Delta\sigma$ represents symmetric behaviour. Clearly to reduce the yield asymmetry, the extrusion speeds must remain low, irrespective of the amount of zinc contained in the alloy. As the extrusion speed is increased, the yield asymmetry declines for all Zn contents. On the positive side, however, for a given value of $\Delta\sigma$, the maximum possible extrusion speed is raised as the zinc content is increased. It should be noted that these higher extrusion speeds may also require very high extrusion pressures which may be unattractive commercially.

2. Fundamentals

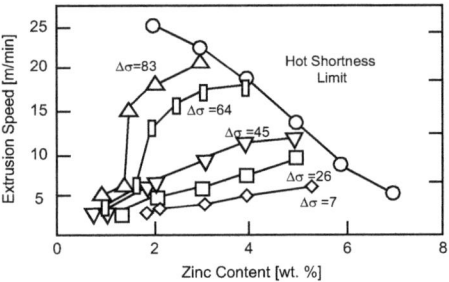

Fig. 2.11. A plot of the extrusion speed vs. zinc content and its effect on the yield asymmetry $\Delta\sigma$ (values in [MPa]) of extruded Mg-Zn alloys [9, 10].

Unfortunately, these results have not been explained up to now in terms of the effect of the Zn addition on the resulting microstructure and the recrystallisation process developed in these alloys. This knowledge is necessary in order to control the resulting mechanical properties of extrusions of these alloys.

One of the most significant results achieved with Zr additions to Mg-Zn binary alloys is that the grain refinement occurring in castings is also translated to enhanced grain refining in the extruded profiles. This increases both the compressive and tensile yield strengths, favouring the former over the latter. Hence, as the grain size is decreased, the yield asymmetry is reduced. In this respect, Doan and Ansel also found that this effect is even stronger for extrusion at low temperatures and very slow speeds, increasing in particular the tensile strength properties. It should also be noted that good compressive yield strength can be achieved in Mg-Zn-Zr alloys, particularly with higher Zn contents. It was noted that the yield asymmetry increases with extrusion speed and that at low speeds (1.52 m/min and less) the values are the same over the zinc range studied (0-7wt. %). However, high extrusion speeds counteracted any grain refining gained by the zirconium additions. On other hand, the microstructures of these alloys in the extruded condition are generally inhomogeneous [9, 49]. Therefore, it is still necessary to improve the extrusion processing of these alloys.

In the case of the rare earth containing alloys, it has been reported that the yield strength of extruded profiles decreases with increasing extrusion speed. The magnitude of the decrease has been related to the amount of rare earth present in the alloy and the progress of

2. Fundamentals

recrystallisation and subsequent grain growth. Similar to the binary Mg-Zn alloys, these results need to be explained on the basis of the recrystallisation process during processing and the texture of the extruded profiles. Moreover the role of Ce-MM as an important alloy addition should be better defined.

2.5 Recrystallisation in magnesium and it alloys

The occurrence of recrystallisation during the thermo-mechanical processing of magnesium alloys is a fundamental process that influences directly the microstructure and mechanical properties of semi-finished products. Therefore it is very important to achieve a better understanding of this phenomenon. In this section, recrystallisation mechanisms reported in Mg alloys will be described.

Recrystallisation is defined as the replacement of a deformed microstructure by a new set of strain-free grains. This process involves the nucleation and subsequent growth of the new grain at the expense of the surrounding deformed grain. The driving force for recrystallisation is the stored strain energy, which is thermally activated by diffusion process. Generally, one distinguishes between a recrystallisation process that occurs during deformation, called dynamic recrystallisation (DRX), or one that takes place subsequent to deformation during an annealing treatment, called static recrystallisation. In a similar distinction, dynamic and static recovery are processes that can also take place during deformation or during annealing, respectively. However, the recovery process is only associated with the restoration of physical properties by rearrangement or annihilation of dislocations without any observable change in the microstructure.

Dynamic recrystallisation is important in many engineering materials including magnesium alloys as it occurs during plastic deformation at temperatures $> 0.5 T_m$, where T_m = melting temperature in °C), i.e. during standard extrusion processes. The critical values of stress and strain at which dynamic recrystallisation is initiated are both material and deformation dependent, e.g. many concentrated alloys and dispersion-strengthened materials do not recrystallise dynamically. Dynamic recrystallisation is beneficial to hot deformation since it not only gives stable flow and good workability to the material, but also reconstitutes the microstructure [73, 74]. Its occurrence can therefore influence the processing window. The flow stress for set-off dynamic recrystallisation decreases with rise of the deformation temperature. Moreover, the dynamically recrystallised grain size is directly correlated with the

2. Fundamentals

flow stress in such a way that with increasing flow stress the grain size decreases. Knowledge of the dynamic recrystallisation characteristics enables the resulting grain size to be adjusted by an appropriate choice of deformation conditions.

The difference between dynamic and static recrystallisation is based on the fact that the dynamic recrystallisation characteristics are decided by the rate of nucleation versus rate of growth under given imposed conditions of temperature and strain rate, whereas static recrystallisation depends on the release of stored energy resulting from hot deformation by thermally activated dislocation processes. The stored energy is provided by accumulations of dislocations formed during work hardening. Highly deformed grains (or highly deformed regions within a grain) have a tendency to nucleate new grains, given the availability of significant misorientations, whereas grains with lower stored energies have a tendency to grow at the expense of their neighbours.

Given that dynamic recrystallisation is the principal process occurring during the extrusion of magnesium alloys, the next section deals with a description of some of its mechanisms. They can be basically differentiated by virtue of the sites where nucleation of new grains takes place and their resulting orientations. The process conditions, principally temperature and strain rate, may also activate a specific recrystallisation mechanism.

2.5.1 Dynamic recrystallisation mechanisms

Although mechanisms of DRX in magnesium alloys are not yet well understood, there are some important references describing its occurrence in magnesium and its alloys [5, 17, 75]. DRX mechanisms can generally be divided into two groups called continuous and discontinuous recrystallisation. A continuous recrystallisation process involves the continuous absorption of dislocations in sub-grain boundaries (low angle boundaries) which eventually results in the formation of high angle boundaries and thus new grains [17]. Discontinuous recrystallisation is a process characterised by the nucleation and growth of new grains via high angle boundary migration (also called primary recrystallisation). It is called discontinuous because the dislocation density in the deformed grains is not reduced homogeneously but discontinuously at a moving grain boundary [76]. In general, during continuous recrystallisation dislocations remain in the recrystallised grains whereas discontinuous recrystallisation removes dislocations through the sweeping action of high angle boundaries. Both mechanisms are applicable to magnesium alloys and a classification in one or other category is made on the basis of the potential nucleation mechanism.

DRX mechanisms have been proposed according to the sites where nucleation of the new recrystallised grains take place: (i) grain boundary nucleation, , (ii) twinning [75], and (iii) particle stimulated nucleation (PSN) [7, 77, 78].

(i) Grain boundary nucleation

Grain boundary nucleation can occur by two different mechanisms: continuous and discontinuous. Independent of these, a necklace-type structure will be developed. This is characterised by fine recrystallised grains located along the original boundaries. A model of this mechanism was first proposed by Ion et. al.[17] for a Mg-0.8wt. % Al alloy and then extended by Galiyev et. al.[5] to a ZK60 alloy. They suggested that three different grain boundary nucleation mechanisms can take place depending on the deformation temperature. Their model is schematically shown in Fig. 2.12 and summarised in the following:

At the relatively low temperature of 150 °C, Fig. 2.12 (a), basal $\langle a \rangle$ slip and twinning systems operate. Basal dislocations accumulate near the grain and twin boundaries and the internal stress in the vicinity of the boundaries exceeds the CRSS for non-basal slip systems. Then, high angle grain boundaries are built by rearrangement of dislocations (a small amount of sub-boundary migration), together with the formation of geometrically necessary dislocations. The orientation of the new grain is slightly different from that of the original grain.

In the intermediate temperature range 200 °C ~ 250 °C, Fig. 2.12 (b), cross-slip of the $\langle a \rangle$ dislocations takes place at regions of high stress concentration, e.g. grain boundaries. After cross-slip, dislocations lying in the non-basal planes are able to climb based on the fact that the stacking fault energies of non-basal planes are higher than that of the basal plane. The dislocation rearrangements brought about by cross-slip and climb generate a low angle grain boundary in the vicinity of the original grain boundary. Finally a high angle boundary is formed as a result of continuous dislocations rearrangement.

At high temperatures in the range 250 °C~450 °C, Fig. 2.12 (c), volume diffusion is possible. Consequently, dislocation climb is more extensive than at lower temperatures. Because of the large amount of dislocation climb, strain localisation occurs at slip lines which cause bulging of the grain boundaries. New grains are formed at the bulges of the original grain boundaries.

Moreover, climb of the dislocation itself leads to the formation of low angle boundaries and moving dislocations are continuously trapped by the low angle boundaries. In this process, the low angle boundaries are gradually converted into high angle boundaries. Due to the progressive increase in grain boundary misorientation in this range of temperatures, this mechanism is also referred to in the literature as rotational dynamic recrystallisation [79].

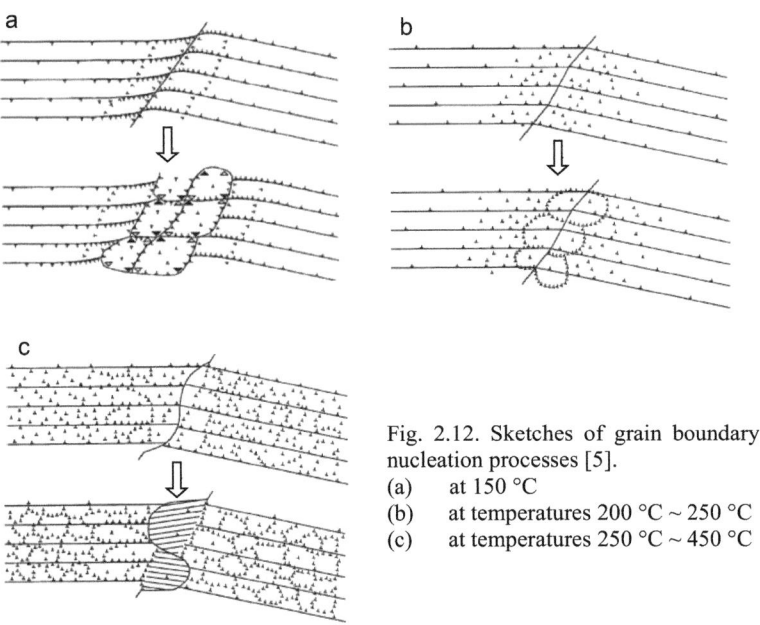

Fig. 2.12. Sketches of grain boundary nucleation processes [5].
(a)   at 150 °C
(b)   at temperatures 200 °C ~ 250 °C
(c)   at temperatures 250 °C ~ 450 °C

(ii) Twinning

DRX based on twinning has been proposed to explain the microstructural evolution during warm and hot uniaxial compression of commercially pure magnesium [80]. Nucleation of DRX at twin-twin and twin-grain intersections has also been observed in hot torsion tests on AZ31 performed at temperatures from 180 °C to 450 °C and strain rates from 0.01 s^{-1} to 1 s^{-1} [75]. These intersections reached a high enough degree of misorientation to activate nucleation of DRX. There is evidence to show that if twinning occurs in the early stages of recrystallisation, it may create the mobile high angle boundary needed for recrystallisation [81]. As twinning produces new orientations that were not present in the initial microstructure, it may play an important part in the development of DRX.

2. Fundamentals

(iii) Particle stimulated nucleation (PSN)

This mechanism consists in the heterogeneous nucleation of at least one grain at the surface of a particle, although there is evidence that multiple nucleation occurs at very large particles [81]. This mechanism occurs in aluminium alloys containing coarse distributions of second phase particles. If the particles are hard and large, they cannot be deformed and geometrically necessary dislocations will generate an inhomogeneous dislocation structure around these particles which thus facilitates PSN [76]. Finely-dispersed particles are less influential for nucleation but they hinder dislocation motion and grain boundary migration, thus recrystallisation can be retarded.

PSN nucleation has been reported in the magnesium alloy WE43 (Mg-4wt. %Y-3wt. %Nd-0.5wt. %Zr)[77, 78]. Nucleation of new grains was observed at second phase particles and often at particles associated with grain boundaries. One important characteristic of PSN is the fact that nuclei surrounding the particle develop a random orientation resulting in a weakening of the recrystallisation texture [7]. Due to its practical interest, PSN remains the subject of current research. However, investigations conducted by Bohlen et. al. [8] on recrystallisation in other magnesium alloys have not shown evidence for PSN.

2.6 Development of texture in extruded magnesium alloys

As a consequence of the effects of deformation and recrystallisation during extrusion, magnesium alloys develop a texture. More as a rule than an exception, grains of a polycrystalline material (considered individually as single crystals) align in some preferred orientation after thermo-mechanical processing, giving rise to a crystallographic texture. If the texture is strongly developed, global behaviour of the polycrystalline material is somewhat similar to that of a single crystal. Thus, the inherent anisotropy of the hcp structure of magnesium (section 2.2) is also carried over to the thermo-mechanically processed polycrystalline material. Specific planes or directions may be strongly oriented with respect to a specific geometric direction in the material, e.g. the extrusion direction of a profile. This dominant orientation can be identified as the texture component of a specific crystallographic orientation. Texture analysis is carried out e.g. using the X-ray diffraction technique. Details on the determination of texture can be found in section 2.8.2.

2. Fundamentals

In the following, the development of texture in extruded magnesium is firstly described on the basis of a single contribution due to deformation. Secondly, the additional contributions due to dynamic and static recrystallisation will be considered.

The development of a preferred orientation resulting from plastic deformation is strongly dependent on the slip and twinning systems available for deformation. The development of texture can be attributed to the operative slip systems and to a greater extent to twinning. Fig. 2.13 shows an example of the changes in orientation of polycrystalline magnesium under compression and tension loading [82]. The inverse pole figures show the crystallographic directions in a direction parallel to the loading direction.

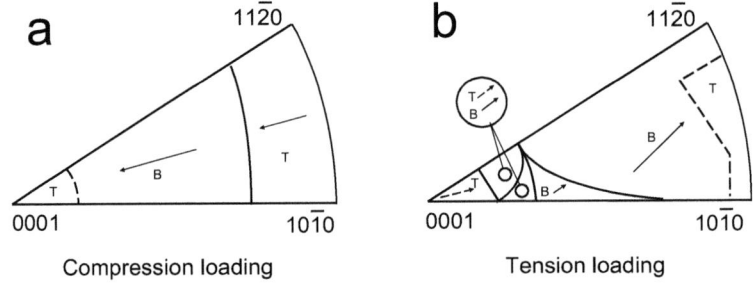

Fig. 2.13. Texture development assuming basal slip (B) and twinning (T) during (a) compression loading and (b) tensile loading [82].

In this example, only slip on the basal system and twinning corresponding to the tensile twin system $\{10\bar{1}2\}\langle10\bar{1}1\rangle$ are assumed. Under compression loading, planes will be oriented to $\langle0001\rangle$ (see Fig. 2.13 (a)), because massive twinning takes place especially in such basal planes aligned perpendicular or within a range of about 70-90 ° to the loading direction. The dashed lines delineate the new orientations and the solid lines the orientations of the original planes. $\langle a \rangle$ slip likewise produces tilting of the c-axis in the compression direction, thus developing a basal texture.

Under tensile loading the occurrence of twinning (compression twins) produces a reorientation of the crystals close to the tension direction by about 86°, see Fig. 2.13 (b). All other orientations are rotated by slip in such a way that the $\langle11\bar{2}0\rangle$ and the $\langle10\bar{1}0\rangle$ prismatic directions become aligned to the tension direction, resulting in a marked double phase texture.

2. Fundamentals

For more complex conditions of deformation, e.g. during extrusion, the development of texture becomes more complicated because of the activation of other additional slip systems and their strain activity dependence. In addition to the deformation conditions, other factors such as chemical composition, the presence of second phases and process temperature are able to influence the activation of slip systems and this may also produce changes in texture. Although the extrusion process is technologically referred to as a multi-axial compression deformation process, the critical deformation zone that determine the resulting orientation of extruded profiles lies in the die exit, at the point where the material is constrained by the die wall. This constraint allows only deformation along the extrusion axis, therefore the material flows only in this direction as in a conventional uniaxial tensile test. Thus, it is considered that the texture of these products is a result of uni-axial deformation.

In the case of magnesium and its alloys with an hcp structure this uniaxial deformation generates a texture in which the {0001} basal planes are parallel to the extrusion axis [83]. Additionally, in the case of round profiles basal planes are radially oriented about this axis. As consequence of basal plane rotation, prismatic planes coincide with the extrusion direction [84]. However, extrusions of magnesium alloys tend to develop a strong $\langle 10\bar{1}0 \rangle$ texture component. A typical example of a so-called $\langle 10\bar{1}0 \rangle$ texture component [85, 86] can be seen in Fig. 2.14 for an extruded round profile of a ZK30 alloy.

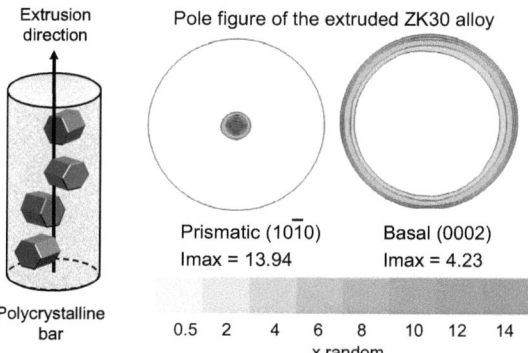

Fig. 2.14. Typical pole figures of round extruded profiles of a commercial ZK30 magnesium alloy parallel to the extrusion direction (from an investigated sample in this work).

2. Fundamentals

This figure shows pole figures for the direction parallel to the extrusion direction, where the prismatic poles $(10\bar{1}0)$ are mainly oriented parallel to this direction and the basal plane (0002) poles are mainly oriented parallel to the transverse direction. The intensity levels registered by texture measurement are represented by gray scale colours in arbitrary units of multiples of random distribution.

Although this prismatic component is very dominant in magnesium extrusions, it is possible that other texture components develop. The type of texture may vary according to the chemical composition and deformation temperature. Fig. 2.15 provides an overview on the influence of the alloying elements on the measured textures in extruded Mg alloys; for example, AZ31 indirectly extruded at 250°C [62] shows the ring type texture in which all crystallographic directions in the basal plane can be laid parallel to the extrusion direction as mentioned above (a). On the other hand, a $\langle 10\bar{1}0 \rangle$ texture component was observed in extruded Mg alloys containing Al [87], Mn [88], or Zn [86] (b), whereas an Mg-1 wt.% Mn alloy extruded at 375°C showed a strong $\langle 11\bar{2}0 \rangle$ component (c) [89] or a double texture component with $\langle 10\bar{1}0 \rangle + \langle 11\bar{2}0 \rangle$ for a extrusion trial at 300°C (d) [88]. Some RE additions produce a $\langle 11\bar{2}1 \rangle$ peak (e) [88, 90].

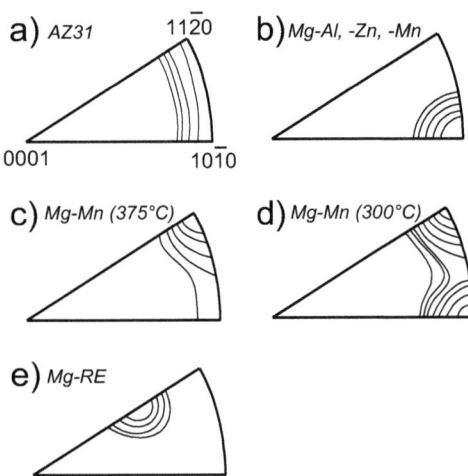

Fig. 2.15. Some examples of textures components in extruded Mg alloys; inverse pole figures in the main deformation direction [85, 90].

2. Fundamentals

The texture of magnesium extrusions can be modified significantly in alloys containing RE additions. The most remarkable effect has been observed in alloys containing Nd or Y, in which more random textures are developed during hot extrusion [7]. This influence may be direct, for example, caused by a change in the activation of slip systems, or indirect, for example, as a result of the temperature increase during extrusion and its effect on the dynamic recrystallisation process involved. Thus, the results shown in Fig. 2.15 could also be explained on the basis of the kind of DRX mechanisms developed during processing and its correlation with the chemical composition.

2.6.1 Contribution of dynamic recrystallisation to the texture development

During thermo-mechanical processing the microstructure is liable to undergo dynamic recrystallisation, (DRX). This mechanism is important because it may be able to affect the orientation of new grains in the resulting microstructure, i.e. the texture and mechanical behaviour of the extruded profiles.

DRX is a strain-induced and thermally-activated mechanism. Therefore, process variables (e.g. temperature or strain rate) could also affect the resulting texture. In the case of the extrusion of magnesium, changes in process parameters such as extrusion speed and ratio and the associated change in temperature during the process can strongly influence the resulting texture.

The texture formed as result of recrystallisation of magnesium alloys has received some attention [91, 92]. This texture is determined primarily by two factors: i) the nucleation of the new grains, and ii) the growth of these grains. Usually the former has a stronger influence on the resulting recrystallisation texture. It is clear that preferential growth may favour some types of nuclei over others, however only orientations which are nucleated have a chance of being part of the final texture. Thus, the above mentioned DRX mechanisms are more concerned with the nucleation than with the growth process. There are basically two possibilities to consider: either the original deformation texture is preserved during recrystallisation, or it is changed considerably. The former relies on the continuous recrystallisation mechanism, which involves recovery, and the latter on e.g. particle stimulated nucleation. Both mechanisms can lead to a weakening of the deformation texture.

2. Fundamentals

The other possibility to modify texture is via secondary recrystallisation. Since a pronounced texture results in a grain boundary character distribution with many low-angle grain boundaries with low mobility, the few grains that are bordered by high-angle grain boundaries can grow and eventually consume the granular microstructure. Therefore, the texture may change during grain growth, both during continuous and discontinuous grain growth, i.e. secondary recrystallisation [76].

During processing, other mechanisms such as dynamic recovery are active. However, recovery does not change the texture for the following reasons. Essentially, recovery competes and also promotes recrystallisation, since it leads to the formation of nuclei for primary recrystallisation. Static and dynamic recoveries are related to a mechanical instability with increase in the temperature. This instability is a consequence of the development of several thermally activated processes, i.e., cross slip of screw dislocations and climb of edge dislocations by which the dislocations unlock and move to interact with other dislocations. Through climb, dislocations can leave their glide planes to arrange in more energetically favourable patterns. Dislocations can mutually annihilate or leave the crystal altogether. Therefore, the recovery mechanism does not contribute to a change in texture [81]. In the next section, the possible influence of static recrystallisation on texture will be described.

2.6.2. Contribution of static recrystallisation to texture development

It is possible that the dynamic recrystallisation process is not completed during deformation; the resulting microstructure will therefore show an inhomogeneous distribution of recrystallised grains coexisting with unrecrystallised grains. The fraction of unrecrystallised grains is able to undergo static recrystallisation if the material is subjected again to high temperatures. Annealing treatments are adequate to investigate possible changes in the deformation texture of materials after thermo-mechanical processing, especially if the material has a considerable fraction of unrecrystallised grains. It is important to mention that besides the fraction of unrecrystallised grains, alloying elements may also influence the development of static recrystallisation. Therefore, annealing treatments on different extruded alloys have been included in this investigation to provide more information on this scarcely researched topic.

In the literature, few reports on the effects of annealing following deformation in magnesium alloys have been mentioned. Pérez-Prado et al [93] reported that secondary recrystallisation

2. Fundamentals

took place in a sheet of the AZ31 alloy produced by extrusion, after a moderate annealing treatment (30 min at 450 °C) and that grains with $\{10\bar{1}2\}$ prismatic planes parallel to the sheet plane grow abnormally.

In the same manner that solute elements affect the activity of different slip systems [12, 25], they may also influence the recrystallisation texture. Since mechanical properties are texture dependent, they could be also texture controlled.

2.7 Yield asymmetry in extruded magnesium alloys

After extrusion, the profiles of magnesium alloys show a yield asymmetry $\Delta\sigma = TYS - CYS$, when a tensile or compressive load is applied to the textured material. A requirement for the commercial application of wrought magnesium alloys is a reduction in the yield asymmetry, i.e. isotropic mechanical behaviour.

The resulting texture may also influence the activation of deformation systems and thus, the mechanical properties, e.g. the yield asymmetry. However, the influence of the grain size on the yield asymmetry can be stronger than that of the texture. Whereas a refined microstructure leads to increased yield strength and ductility together with the possibility to reduce the yield asymmetry; coarse grains promote profuse twinning and at the same time enhance the yield asymmetry.

2.7.1 Yield asymmetry in terms of the relationship grain size – twinning activation

The effect of twinning during room temperature mechanical testing of extruded magnesium alloys can be illustrated by considering the material shown in Fig. 2.14, in which basal planes are strongly oriented parallel to the extrusion direction. If a compression load is applied parallel to the extrusion direction (contraction perpendicular to the c-axis), then $\{10\bar{1}2\}$ tensile twins are readily activated leading immediately to plastic deformation.

2. Fundamentals

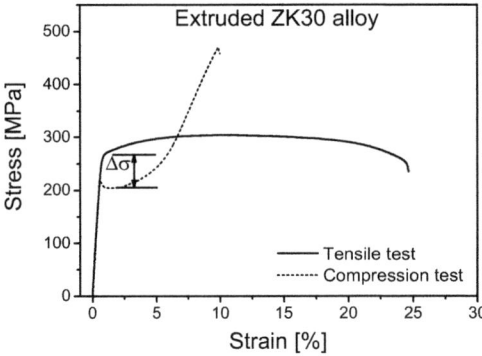

Fig. 2.16. Engineering stress-strain flow curves of an extruded ZK30 sample loaded parallel to the extrusion direction (from an investigated sample in this work).

On other hand, if a tensile load is applied parallel to the extrusion direction (extension perpendicular to the c-axis), then $\{10\bar{1}1\}$ compression twins will be activated. However, the shear stress necessary to activate this twinning mode is higher than that for tensile twins. This results in higher tensile yield strength necessary to start plastic deformation parallel to the extrusion direction. This effect is clearly observed in Fig. 2.16 which shows the stress-strain curves from tension and compression tests on the same extruded ZK30 alloy.

The plastic behaviour observed immediately after yielding also shows a strong influence of twinning. On one hand, during compression test the yield strength shows a maximum followed by an apparent drop of the stress. The fact that $\{10\bar{1}2\}$ tensile twins are easily activated leads to a rapid development of twins. The yield drop observed is a result of the fact that profuse twinning provides an extension which is larger than the length change imposed by the mechanical testing machine [76]. Consequently, the specimen is partially unloaded for a short time. The twins formed in the early stages of deformation tend to act as barriers to the propagation of dislocations, which causes a rapid increase in the hardening rate. Thus, the flow curves tend to develop an S-shape form. This is identified as a sigmoidal hardening by twin-dominated deformation and is typical for extruded magnesium alloys [94].

On the other hand during the tension test, twinning occurs but it does not dominate during plastic deformation. Instead, other slip systems can be activated at higher stresses leading to a parabolic hardening characteristic of slip-dominated deformation, also typical behaviour of conventional magnesium extrusions in tension [94, 95].

2. Fundamentals

Twinning is the principal reason for the yield asymmetry in single crystals as well as in extruded round profiles. However it has been shown that the activation of twinning also depends strongly on the grain size in extruded magnesium alloys [96]. Generally, the grain size dependence of the activation of deformation mechanisms in polycrystals is described by the Hall–Petch relationship, [22]

$$\sigma = \sigma_0 + k \cdot d^{-1/2} \qquad (2.1)$$

where σ is the yield strength, σ_0 is the "friction stress" representing the overall resistance of the crystal lattice to dislocation movement, k is the "locking parameter" which measures the relative hardening contribution of the grain boundaries and d is the grain diameter. This equation is based on the concept that grain boundaries act as barriers to dislocation motion and that dislocations will begin to pile up at the grain boundaries. Dislocations at the tip of the pile-up must experience some critical shear stress to continue slip over the grain boundary barrier. This critical shear stress of the possible activated slip modes is directly related to the factor k. The factor k is represented by the slope of the straight line that is obtained when σ is plotted against $d^{-1/2}$, which is roughly independent of the temperature. The term σ_0 is the intercept along the ordinate in the same plot. It is interpreted as the friction stress needed to move unlocked dislocations along the slip plane.

However, to what extent this relation can also be applied to deformation twinning in magnesium has only been studied briefly [31]. For body centred cubic (bcc) metals such a description has been shown [29]. By using textured profiles, the contribution of twinning can be maximised in compression whereas it can be minimised in tension. Thus, any effect of the grain size or the alloying element content on twinning will be magnified and can be studied by a Hall-Petch analysis, using the tension/compression yield asymmetry ($\Delta\sigma$). Although it is not possible to determine the value of the k parameter for deformed and textured polycrystalline materials (only if the texture remain the same), the Hall-Petch relationship is an important graphical tool that shows not only the effect of a reduction in grain size on the twinning activity in extruded magnesium alloys, but also the effect of the alloy element addition on the twinning activity [3, 96].

This relationship will be used in the present investigation in order to discuss the separate effects of grain refinement and alloying element additions on the yield strength and yield asymmetry of the extruded profiles. Additionally the possibility to integrate the effect of the texture in a Hall-Petch plot will be discussed.

2.8 Experimental fundamentals

2.8.1 Extrusion methods and process parameters

Extrusion methods are classified as (i) direct, (ii) indirect and (iii) hydrostatic. The most common method in industrial use is direct extrusion, because the tooling is not complicated and the process, while not used for high-precision applications, provides an acceptable product. Nevertheless, indirect extrusion is also a widely used process. It has been demonstrated experimentally that hydrostatic extrusion is particularly advantageous for magnesium alloys from a technological point of view [62, 66].

(i) Direct and (ii) Indirect extrusion [61, 69]

For the direct process, the metal billet is placed in a container and driven through the die by the ram. A dummy block is placed at the end of the ram in contact with the billet; see Fig. 2.17 (a). In the indirect extrusion process, a hollow ram carries the die, while the other end of the container is closed with a plate. The ram containing the die is kept stationary, and the container with the billet is mobile, see Fig. 2.17 (b). Because there is no relative motion between the wall of the container and the billet in the indirect extrusion, the frictional forces are lower and the power required for extrusion is less than for direct extrusion. Friction occurs solely between the container and the die. In this method the profile size is limited by the cross sectional area of the hole in the hollow ram and this in turn by the load on the hollow ram.

The advantages of indirect extrusion with respect to the direct method are the lower load required and the development of a more uniform flow pattern because of the absence of relative motion between the billet and the container. Moreover, heat is not produced by friction between the billet and the container. Consequently, there is a considerable reduction in the temperature. These process conditions make indirect extrusion a very stable process with the possibility to achieve uniform deformation of the complete billet cross section.

(iii) Hydrostatic extrusion [97, 98]

For the hydrostatic process, the billet in the container is surrounded by a fluid, a so called hydrostatic medium. On the stem side the container is sealed with the ram and on the die side with the billet so that the ram can compress the hydrostatic medium without coming into contact with the billet, Fig. 2.17 (c). The most important feature is that the hydrostatic pressure is applied to the medium leading to a homogeneous distribution of pressure over the whole surface of the billet. Thus, higher pressures than those available in the other methods can be applied. Moreover, the influence of friction is eliminated because the ram and the container do not touch the billet directly and the die is the only possible place where friction may appear. The lack of friction between billet and container also serves to reduce the extrusion force required. This allows the use of higher extrusion ratios and therefore faster production speeds. Since the heat developed by deformation is not excessive, the limitation of the process is dictated by the ability of the alloy to withstand plastic deformation at low temperatures. Lower process temperatures allow finer grain sizes to be obtained.

In contrast to the other processes, each billet has to be tapered to match the die angle in order to form a seal at the start of extrusion. One disadvantage is that the production of complex profiles is restricted due to technical problems with the development of adequate dies.

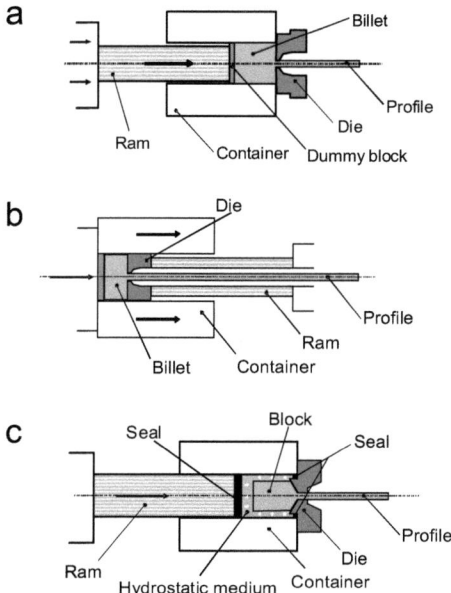

Fig. 2.17. Diagram of the direct (a), indirect (b) and hydrostatic (c) extrusion processes [99].

Process parameters

The principal process parameters which may be controlled during extrusion with these three different methods are (1) the extrusion ratio, (2) the billet temperature, (3) the ram or product speed, and (4) the frictional conditions at the die and container wall. The first three are most usually varied in order to control the extrudability of the material and the resulting conditions of the profile. The influence of the latter is very significant for the direct method, negligible for the hydrostatic process and almost eliminated for the indirect press machines through the use of lubricants, which reduce the friction between the die and the container, as well as between the die and the flowing material.

The extrusion ratio, $\varphi = A_0/A_f$, where A_0 is the transversal area of the container and A_f is the transversal area of the profile, is generally fixed due to specific requirements of the customer's product and the available container size; therefore, it is not readily controllable. The billet temperature and the ram-product speed (speed of deformation) are related to each other. If we consider that the result of an increase in the speed of deformation is an increase in the temperature, then limiting factors that determine whether or not extrusion may be successfully conducted are reduced to two parameters.

The first limit is the extrusion pressure, which must not exceed the capacity of the extrusion press. This parameter is better controlled in hydrostatic extrusion, due to the higher capacities of these presses (up to 40 MN). The second one is the maximum temperature that the material can tolerate during processing before the surface quality deteriorates or hot cracking occurs.

2.8.2 Pole figure goniometer and texture analysis [21, 100]

A pole-figure goniometer using reflection geometry and monochromatic X-rays Cu Kα radiation can be used for quantitative determination of textures. In this device, the source and counter are arranged in a fixed geometry, depending only on the Bragg angle of the investigated crystallographic plane. The Bragg angle refers to a reflection condition if Bragg's law is obeyed:

$$n\lambda = 2d \sin \theta \qquad (4.1)$$

where λ is wavelength; d is the spacing of the reflecting planes; θ is the angle of incidence and reflection and n is the order of diffraction. The sample is mounted on a holder which can be rotated around two mutually perpendicular axes to orient the specimen in any position with respect to the incident X-ray beam. The goniometer moves the detector with respect to the X-ray beam by two rotations, Φ and χ (see definitions in Fig. 2.18). The χ circle is generally symmetrical between the incoming and diffracted beam (positioned at an angle θ). The 2θ and

2. Fundamentals

ω axis coincide. The nomenclature Φ, χ and θ is standard in single-crystal diffractometry and marked on most instruments. Stepper motors, controlled by a personal computer, enable one to obtain any arbitrary angular position on the three axes 2θ, χ and Φ (within a certain range to avoid mechanical collisions). The axe ω sets only the detector to the proper Bragg angle, 2θ, of the diffraction peak of interest.

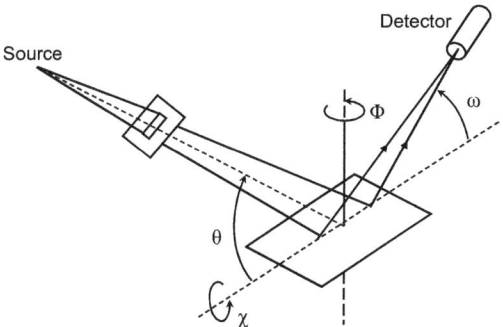

Fig. 2.18. Ray path and sample rotation in an X-ray texture goniometer and definition of the instrument angles [100].

The movement of the specimen unveils the spatial orientation of the respective poles {hkl}. In a stereographic projection, the measured intensity distribution generates the {hkl} pole figure. A pole figure shows the distribution of a selected crystallographic direction relative to certain directions in the specimen (see Fig. 2.14). In the case of magnesium, typical pole figures of interest are { 0002 }, {10 $\bar{1}$ 0 }, {11 $\bar{2}$ 0 }, {10 $\bar{1}$ 1}, {10 $\bar{1}$ 2 }, and {10 $\bar{1}$ 3 }.

Another representation is the inverse pole figure, which is more adequate for showing condensed information on pole figures. This presentation shows the distribution of orientations in a selected direction of the specimen. The projection plane of an inverse pole figure is therefore a standard projection of the crystal, of which only the unit stereographic triangle needs to be shown. Fig. 2.19 shows the stereographic projection of a standard representation of an inverse pole figure in the hcp lattice structure. Considering the symmetry of the extruded profiles, inverse pole figures contain most of the relevant information related to the extrusion direction, thus, this presentation will be used in the results.

2. Fundamentals

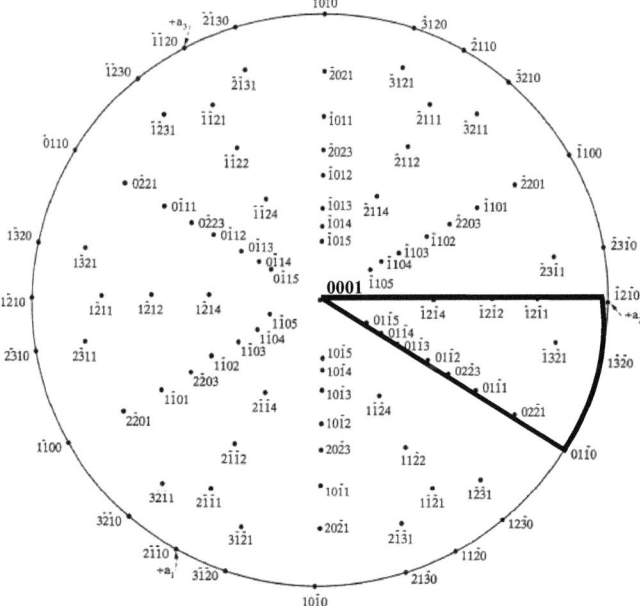

Fig. 2.19. Standard (0001) stereographic projection of an hcp crystal structure showing a diagrammatic illustration of the inverse pole figure (triangle) used in this study for the representation of textures in extruded profiles [21].

3. Experimental details

3.1 Investigated alloys

Based on the information compiled in fundamentals, several magnesium-zinc based alloys were selected for this investigation. Considering that the aims of the work are focussed on the effects of the alloying elements on magnesium extrusions and that zinc is the base element, a basic group consisting of pure magnesium of commercial purity (99.94 wt. % see impurity content in Table 3.1) and four binary Mg-Zn alloys (identified as Z-series) was incorporated.

Table 3.1: Impurities in the commercially pure Mg used in this work (all in wt. %)

Mg	Fe	Si	Ni	Cu	Al	Fe	Mn
99.94	0.0231	0.0087	<0.0002	0.00025	0.0096	<0.1	<0.15

The nominal compositional range of the binary alloys is from 1-4 wt. % Zn (called Z1, Z2, Z3 and Z4). They represent the usual Zn-contents of commercial wrought magnesium alloys. The chemical compositions of the alloys used in this work are summarised in Table. 3.2. The importance of this basis group is that it will define exactly the sole effect of the Zn addition on magnesium during and after extrusion. Moreover, it is necessary to make a progressive contrast with the rest of alloys containing additional alloying elements.

The second group of alloys corresponds to a magnesium-zinc-zirconium alloy called ZK-series alloys, see Table 3.2. These alloys differ only in respect to the addition of approx. 0.5 wt. % Zr, which is an important addition for the Mg-Zn based wrought alloys, principally due to its grain refinement effect. A Mg-Zr master alloy containing 33 wt. % Zr (Zirmax^{TM}) was used for Zr additions. These alloys were called ZK10, ZK20, ZK30 and ZK40. In addition, a commercial ZK60 alloy was included to this series, although, ZK30 and ZK40 can also be considered as commercial wrought alloys. This group was selected based on their good adaptation to extrusion process as well as their good mechanical properties after extrusion. The compositions of these alloys allow a separation of the effect of a sole Zr addition on the binary Mg-Zn alloys. Commercial alloys were included because it is still necessary to understand the reasons for the inhomogeneous microstructures observed in extrusions of these alloys. This knowledge is required in order to improve the extrusions of this group.

3. Experimental details

Table 3.2: Nominal composition of cast billets (all in wt. %, Mg balance). Zn is the only element which is progressively changed.

Alloy		Zn	Zr	RE**
–	Mg*	–	–	–
Z – series	Z1	1	–	–
	Z2	2	–	–
	Z3	3	–	–
	Z4	3.6	–	–
ZK – series	ZK10	1	0.4	–
	ZK20	2	0.5	–
	ZK30	3	0.6	–
	ZK40	3.5	0.6	–
	ZK60	5	0.3	–
ZEK - series	ZEK100	1	0.4	0.8
	ZEK200	2	0.6	0.8
	ZEK300	2.8	0.6	0.8

* Impurities in Table 3.2
** Cerium – Mischmetal

On the other hand, based on their excellent adaptability to the extrusion process, improvements in the resulting mechanical properties and their very high potential for structural applications, alloys of the Mg-Zn-Zr-RE system were incorporated in this investigation. Ce-Mischmetal (Ce-MM) was selected as rare earth addition due to its greater chance to be applied for the development of new commercial magnesium alloys, in the sense of its lower price in comparison with that of pure RE. It is also expected that Ce-MM additions could provide microstructural stability during extrusion of these alloys. For these reasons and following a logical sequence in the chemical compositions with the other group alloy, three ZK alloys were cast with additions of Ce-MM, called ZEK100, ZEK200 and ZEK300. Ce-MM was purchased in the form of small tablets and its chemical composition is shown in e Table 3.3. The constant contents of Ce-MM allow separation of the effect of this addition on the ZK-series alloys.

Table 3.3: Chemical composition of Ce – Mischmetal used in this work (all in wt. %)

Ce	La	Pr	Nd	Si	Fe	Mn	Ni	Cu	Others
57.4	27.3	5.1	1.9	<0.002	<0.1	<0.15	<0.001	<0.002	<1.0

3. Experimental details

3.1.1 Castings

Billets of all the manufactured alloys showed in Table 3.2 were produced by gravity casting. They were melted in a steel crucible (with a maximum capacity of 16 litres) using a resistance furnace. Alloying element additions were added to the molten Mg in the crucible at temperatures between 740 and 760 °C, depending on the alloy. Casting was performed under a protective atmosphere mixture of argon (68.2 litre/hour) and SF_6 (11.71 litre/hour). The melt was held for 60 min under stirring to make sure that the alloying elements were homogeneously dissolved. The alloys were cast into cylindrical steel containers of 100 mm in diameter and 410 mm in length. The weight of the cast billets was about 8 kg. The chemical compositions of the cast alloys shown in Tables 3.1 and 3.2 were determined by spectroscopic analysis using OES Spectrolab™ M equipment. In agreement with the principal alloying element contents and their corresponding binary phase diagrams, a heat-treatment to homogenise the as-cast microstructure at 350 °C for 15 hours was carried out for all the cast billets. At this temperature the principal alloying element additions show a region of solid solution with magnesium.

3.1.2 Characterisation of the cast billets

For microstructural characterisation of the homogenised cast billets, slices of ~2 cm corresponding to both ends of the cast billets were cut, see Fig. 3.1. With this procedure destruction of the extrusion billet is avoided; moreover, the quality of the homogenisation heat treatment is observable at both ends where maximum differences in grain size are obtained after casting (usually the grain size at the bottom is coarser than at the top). From these slices, samples from the centre, edge and centre-edge (shown as A, B, C) were prepared to observe the microstructure and determine the grain size.

3. Experimental details

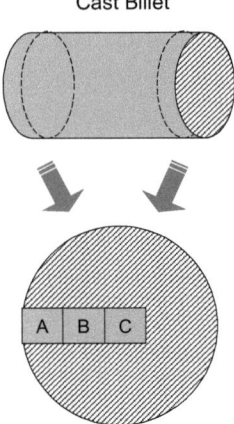

Fig. 3.1 Slices of the cast billets for characterisation. A, B and C indicates the position of samples in the edge, edge-centre and centre of the slide.

3.2 Process

3.2.1 Indirect extrusion

The indirect extrusion trials were performed using a press machine with a maximum press capacity of 8 MN at the Extrusion Research and Development Centre of the Technical University of Berlin, see details in Fig. 3.2.

Specifications:

Construction:
Horizontal

Ram speeds range:
0 – 80 [mm/sec]

Diameters of the container:
85, 95, 110, 125, 140 [mm]

Maximum length of the billet:
500 [mm]

Fig. 3.2. The indirect extrusion press in the Extrusion Research & Development Centre (ERDC) of the Technical University of Berlin.

3. Experimental details

In this press machine, the die is fitted with a thermocouple, see Fig. 3.3. This allows the profile temperature to be measured in situ during the extrusion process.

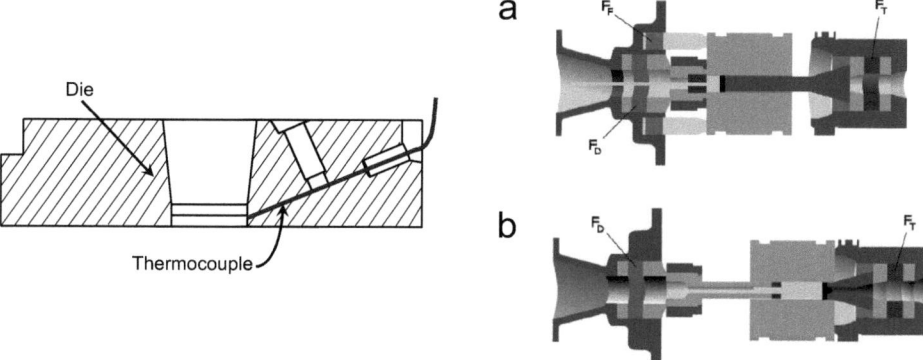

Fig. 3.3. Thermocouple tip at the bearing surface [64].

Fig. 3.4. Configuration of the press during (a) direct extrusion and (b) indirect extrusion. F_F = friction force, F_D = die force and F_T = total press force [64].

Additionally, load cells have been fitted to this press to record the total force F_T, the friction force F_F and the die force F_D ($F_T = F_F + F_D$). This unit can be modified to perform direct and indirect extrusion. Fig. 3.4 shows schematically the press configurations indicating the positions where the various forces are recorded, whereby the friction force F_F only occurs in direct extrusion. Thus, for indirect extrusion $F_T = F_D$.

The cast billets were machined into cylindrical forms with diameters of 93 mm and lengths of ~ 300 mm; see Fig. 3.5. Prior to indirect extrusion, the billets and the die were coated with a combination of beeswax and graphite as lubricant. They were then preheated in an induction furnace to 300 °C. This temperature is considered to be a typical process temperature for the extrusion of magnesium alloys [34, 61, 64] and this was the initial extrusion temperature used for all alloys in this work. The die was also preheated to 300 °C in the same induction furnace.

3. Experimental details

Fig. 3.5. Cylindrical billets for indirect extrusion.

The diameter of the billet is smaller than that of the container (95 mm) because expansion of the billet is expected after heating. Regardless of whether the billet expands to the walls of the container or not, the material will be compressed by the ram until the material completely fills the container. Thus, the container diameter should always be considered as the initial diameter. Taking this and the diameter of the resulting round profile of 17 mm into account, an extrusion ratio φ of 30 is fixed for all the trials. The extrusion rate was varied in a stepwise manner during the extrusion trials in order to achieve profile speeds of 1, 5 and 10 [m/min], see Fig. 3.6. A good approximation for the overall strain rate $\dot{\varepsilon}_t$ during extrusion can be made using the relationship:

$$\dot{\varepsilon}_t = \frac{6\upsilon D_b^2 \ln\varphi \tan\alpha}{D_b^3 - D_e^3} \tag{3.1}$$

where υ is ram speed, φ is the extrusion ratio, D_b is the billet diameter, D_e is the profile diameter considering a round shape rod and α is a semi-die angle, which is considered to have a constant value of 45°. More details on this calculation can be found in [22]. For the profile speeds used this gives mean strain rates of $\dot{\varepsilon}_t$ = 0.11 s^{-1}, 0.6 s^{-1} and 1.21 s^{-1}, respectively. After processing, all profiles were air-cooled. Pure magnesium billets were extruded at only two profile speeds of 2 and 10 (m/min) corresponding to mean strain rates of $\dot{\varepsilon}_t$ = 0.22 s^{-1} and 1.21 s^{-1}.

3. Experimental details

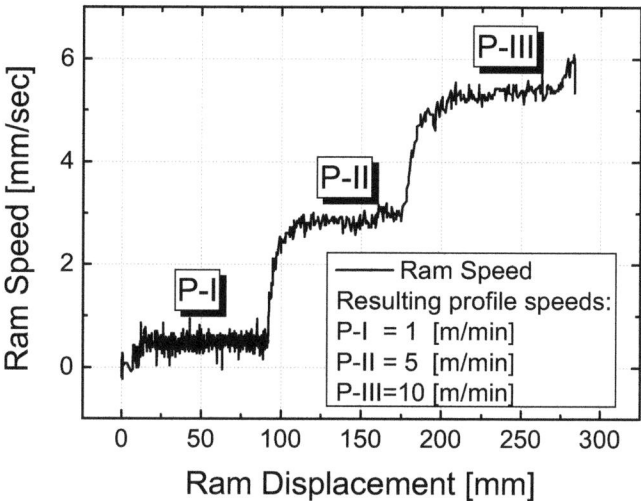

Fig. 3.6 Stepwise variation of extrusion rate during indirect extrusion and the resulting profile speeds.

By mean of computer-aided measuring systems the total force (for convenience called only extrusion force) and the profile temperature (product outlet temperature in the bearing face of the die) could be determined in relation to the ram displacement.

3.3 Annealing treatment after extrusion

An annealing heat treatment of 1 hour at 400 °C was carried out for some indirectly extruded alloys of the ZK and ZEK-series. The aim of this heat treatment was to analyse the progress of static recrystallisation in extruded profiles. The resulting inhomogeneous microstructural condition of these alloys and scarce information on static recrystallisation in these alloys made them suitable candidates for this analysis. For these experiments a circulating air oven was used. A protective atmosphere was not necessary because any oxide layer formed on the sample surfaces is removed during subsequent metallographic preparation. Microstructural characterisation and texture measurements were performed on the annealed samples.

3.4 Characterisation of the extruded profiles

For the evaluation and correlation of the microstructure and mechanical properties of the extruded profiles, metallographic analysis, texture measurements and mechanical testing were carried out. The lengths of the indirectly extruded profiles corresponding to each extrusion speed were between 2.5 and 3 meters. For better representation of the extruded material at

3. Experimental details

each speed, all the samples for characterisation were taken from between the 2nd and 3rd meter of the profile. For mechanical testing, samples were machined parallel to the extrusion direction and at least five tensile and five compressive samples were tested.

For microstructural analysis the extruded profiles were cut transversally and ground to provide samples with a surface perpendicular to the extrusion direction, see Fig. 3.9. These surfaces were prepared to study the microstructure and determine the grain size.

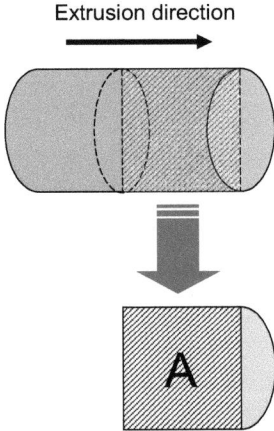

Fig. 3.9. Orientation of the samples used for characterisation of extruded profiles (A) indicates the selected surface.

For texture measurements, transverse slices of the extruded profiles were taken, see Fig. 3.10. The transversal surface of these samples has a diameter of 17 mm and a thickness of ~5 mm. For microstructures with coarse grain sizes (40 - 80 µm), four samples were prepared.

3. Experimental details

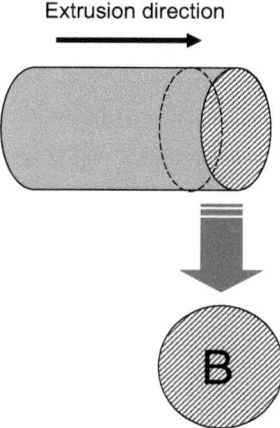

Fig. 3.10. Orientation of the samples used for texture measurements on extruded profiles. (B) indicates the selected surface.

3.5 Metallographic procedures

A standard metallographic procedure was used for microstructural observation of magnesium alloys [101]. This procedure includes preparation of samples destined for texture measurements and details are described in the following.

Cutting and grinding

Samples of the cast material and extruded profiles were carefully cut with a diamond cutting disc at low velocity in a normal water-cooled cutting machine. To eliminate twin formation during cutting, the samples were ground using coarse SiC papers (800 grit) until the selected surface was completely flat. Further grinding was carried out using successively finer SiC papers (1000, 1200 and 2500 grit).

Polishing

After grinding, the samples were polished to remove scratches created by grinding on a porous cloth of Neoprene (MD-Chem, Struers™) using an oxide polishing suspension (OP-S, Struers™) of Al_2O_3 (0.05 µm) and distilled water. Ethanol was used in the last steps of polishing instead of water in order to avoid oxidation. Finally the samples were rinsed and cleaned in an ultrasonic-bath with ethanol.

3. Experimental details

Etching

The polished samples were chemically etched with a solution of picric acid (150 ml of ethanol, 40 ml distilled water, 6.5 ml acetic acid, 30-40 g picric acid (98%)) [101]. This picric acid solution produces clear contrast of the grain boundaries by selectively etching grains depending on their crystallographic orientation. Samples for scanning electron microscopy (SEM) were not etched.

Electropolishing

Samples assigned for SEM observation and texture measurements were additionally subjected to electro-polishing. This procedure allows flat surfaces free of mechanical twins to be obtained. Fig. 3.11 shows the LectroPol-5 (Struers™) electro-polishing device used in this work. It consists of a polishing unit, a cooling system and a control unit. The polishing unit has a container for the electrolyte, a base on the top acting as a sampler holder and an anode which pushes the sample to the base and conducts the current to the sample. The temperature of the electrolyte in the container is controlled by the cooling system. During polishing, the electrolyte is pumped onto the surface of the sample at a specific flow rate and voltage. The AC2 commercial solution (Struers™) was used as electrolyte. Optimum conditions for polishing were achieved with an electrolyte temperature of -25 °C and a constant voltage of 30 V for 7 minutes.

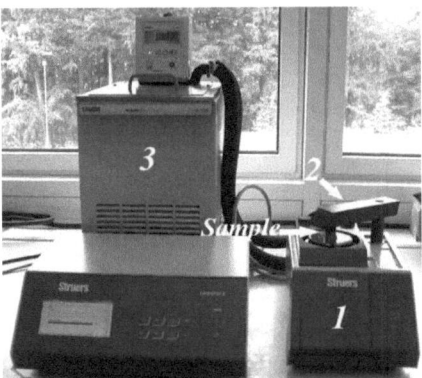

Fig. 3.11. Electro-polishing device. 1) Polishing unit with the electrolyte container; 2) anode and 3) cooling system.

3.6 Microstructural characterisation

Microstructural analysis was performed using optical microscopy and scanning electron microscopy (SEM). Optical microscopy observations were made with a Leica™ DMLM light

3. Experimental details

optical microscope able to achieve magnifications from 1.6 to 1000-fold. Micrographs were recorded using a Color View camera DC500 connected to the microscope and were evaluated using "analySIS" pro version 5.0 software-image processing (Aquinto™). This software allows the average grain size and its distribution to be determined by a linear intercept statistical method in accordance with DIN 50600 [102]. In this method, a grid of five or six parallel lines lying horizontally and vertically to the micrograph is set up in order to manually delineate the grain boundaries. The outcome of this is the calculation of the average surface of the grains so that the average grain diameter (grain size) can be determined. During grain size determination, large and long unrecrystallised grains are considered and counted as normal grains. However, the statistical discrimination of these few grains shown in the micrograph means that they do not represent a measurable influence in the determination of the grain size. Three micrographs from different positions for a single sample were used. In agreement with the norm, it is necessary to count at least 50 grains in each micrograph to assure a valid determination of the grain size. Errors were determined from the standard deviations of the data collected. Additionally, the accuracy of magnification in the microscope was considered to be in the range ± 1 µm.

The morphology and the chemical compositions of the intermetallic secondary phases present in the alloys were analysed using a field emission gun scanning electron microscope (FEG-SEM) ZEISS™ Ultra 55, with an accelerating voltage of 15 or 20 kV. The microscope is also equipped with energy–dispersive X–ray spectroscopy (EDX) equipment. The chemical composition of the particles present in the alloys before and after the extrusion process was determined. The purpose is to know their possible development and influence during processing. The EDX results were based on a considerable number of sample analyses.

3.7 Mechanical testing

For tension and compression tests, a universal testing machine Zwick™ Z050 with a maximum applied force of 50 kN was used. Tensile samples were machined according to standard DIN 50125 with a gauge length of 30 mm, 6 mm diameter and screw heads on the two-ends [103]. Cylindrical compression samples were machined with lengths of 16.5 mm and diameters of 11 mm in accordance with DIN 50106 [104]. In both cases, at least three identical samples were tested. The strain rate for all tests was constant with 10^{-3} s^{-1}. Stress-strain flow diagrams were collected using extensometers.

3. Experimental details

Yield strength values ($\sigma_{0.2}$), called tensile yield strength (TYS) and compressive yield strength (CYS), were determined. The yield strength $\sigma_{0.2}$ was taken as the 0.2% proof stress or as the lower compressive yield strength σ_{lcys}, where applicable. The elongation was determined through strain to fracture (%).

3.8 Texture measurements [100, 105]

Quantitative texture measurements were made with a Panalytical™ pole-figure goniometer using reflection geometry and monochromatic Cu-Kα X-rays. During measurements the surface of the sample lies parallel to the surface of the specimen holder, thus the extrusion direction is parallel to the angle Φ, see Fig. 2.17.

In the case of magnesium, the $\{0002\}$, $\{10\bar{1}0\}$, $\{11\bar{2}0\}$, $\{10\bar{1}1\}$, $\{10\bar{1}2\}$, and $\{10\bar{1}3\}$ pole figures were determined. They were used to calculate the orientation distribution function (ODF) and perform subsequent corrections to the measured pole figures including background defocusing. To perform calculations of the pole figures, a method based on the Williams-Imhof-Matthies-Vinel (WIMV) was used. The ODF function was also used to calculate the inverse pole figures. Only this representation of texture is shown in this work.

The measurement parameters were varied depending principally on the resulting grain size; several samples were necessary for coarse grained microstructures (20 – 80 µm). The X-ray source was adjusted to 2 x 1 mm by a crossed slit collimator and the detector was preceded by Soller slits. The complete pole figures were measured by scans of six seconds at each location in 5 ° tilt steps from 0 to 70 ° on χ and azimuthal steps of 5 ° over the entire 360 ° on Φ. Before the pole figures were collected, standard θ-2θ scans were run in order to obtain the exact positions of the Bragg peaks for each alloy.

Additionally to these measurements, a couple of extruded profiles were selected to analyse local information on microstructure and crystal orientation by means of EBSD measurements. For this purpose, a EDAX/TSL™ EBSD system adapted to the same scanning electron microscope previously mentioned was used. Orientation image mapping (OIM) was conducted in 0.3 µm steps with 16 mA and 15 kV of accelerating power. The analysed surface corresponds to that of Fig. 3.9. Samples preparation was similar to that used for texture measurements.

4. Results

4.1 Characterisation of the homogenised billets

In the following, the microstructures and grain sizes of the cast billets after the homogenisation treatment as determined by optical microscopy will be described. Figs. 4.1, 4.2, 4.3 and 4.4 show the microstructures of the homogenised billets of commercial purity Mg, the Z-series alloys, the ZK-series alloys and the ZEK-series alloys. The results of SEM studies on the nature of the particles present in the various alloys will then be described.

Microstructure of commercial purity Mg

The microstructure of commercial purity magnesium in the homogenised billet consists of very coarse grains, see Fig. 4.1, especially in the central region of the billet.

Mg (1200 ± 500 µm)

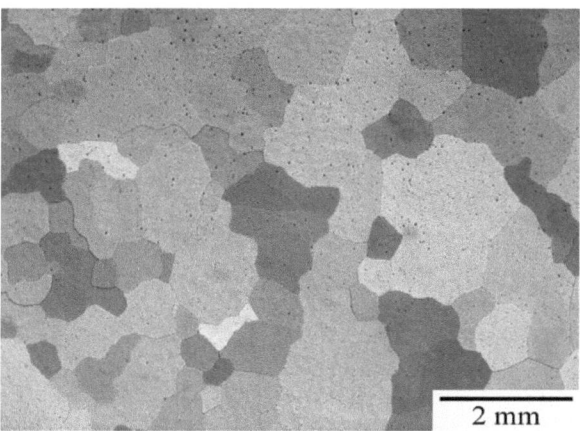

Fig. 4.1. Commercially pure Mg after a homogenisation heat treatment.

4. Results

Microstructures of the Z-series alloys

Additions of zinc lead to significant grain refinement. Figs. 4.2 (a)-(d) show the effect of Zn additions in solid solution on the microstructure. The microstructures consist of homogeneously distributed equiaxed grains.

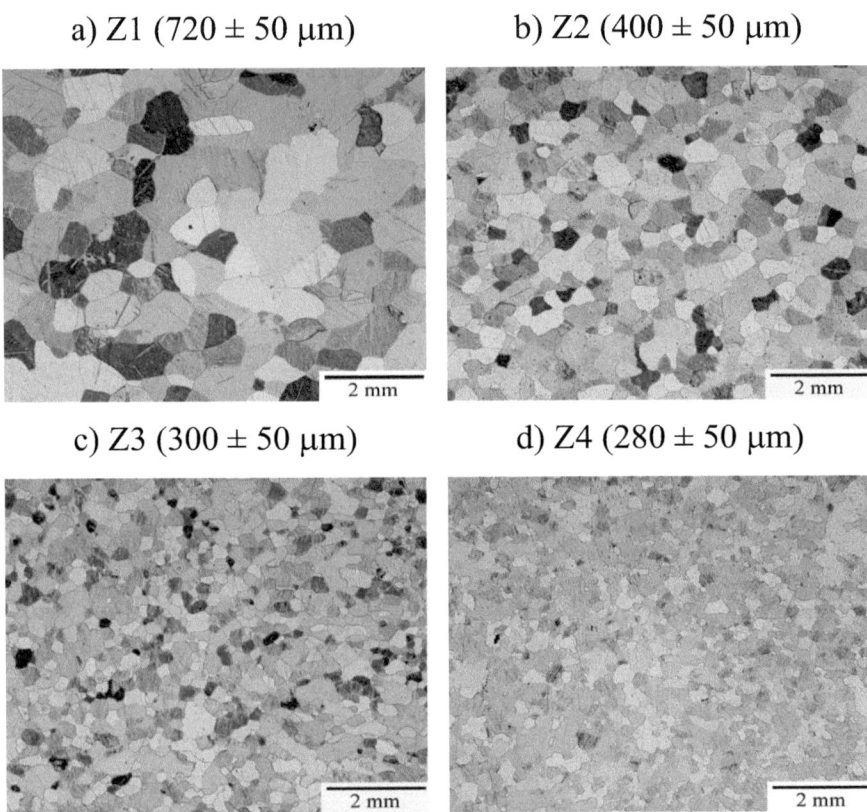

Fig. 4.2 Microstructures of the homogenised Z-series alloys.

Microstructures of the ZK-series alloys

Even more impressive is the effect of Zr additions to the Mg-Zn alloys, where a further significant reduction in the average grain size is obtained, see Fig. 4.3 (a)-(e). With increasing Zn content the grain sizes were also somewhat reduced. The microstructures of this group of alloys (ZK-series) also consist of homogeneously distributed, equiaxed, globular grains.

4. Results

a) ZK10 (150 ± 40 μm) b) ZK20 (140 ± 20 μm)

c) ZK30 (135 ± 20 μm) d) ZK40 (100 ± 10 μm)

e) ZK60 (100 ± 20 μm)

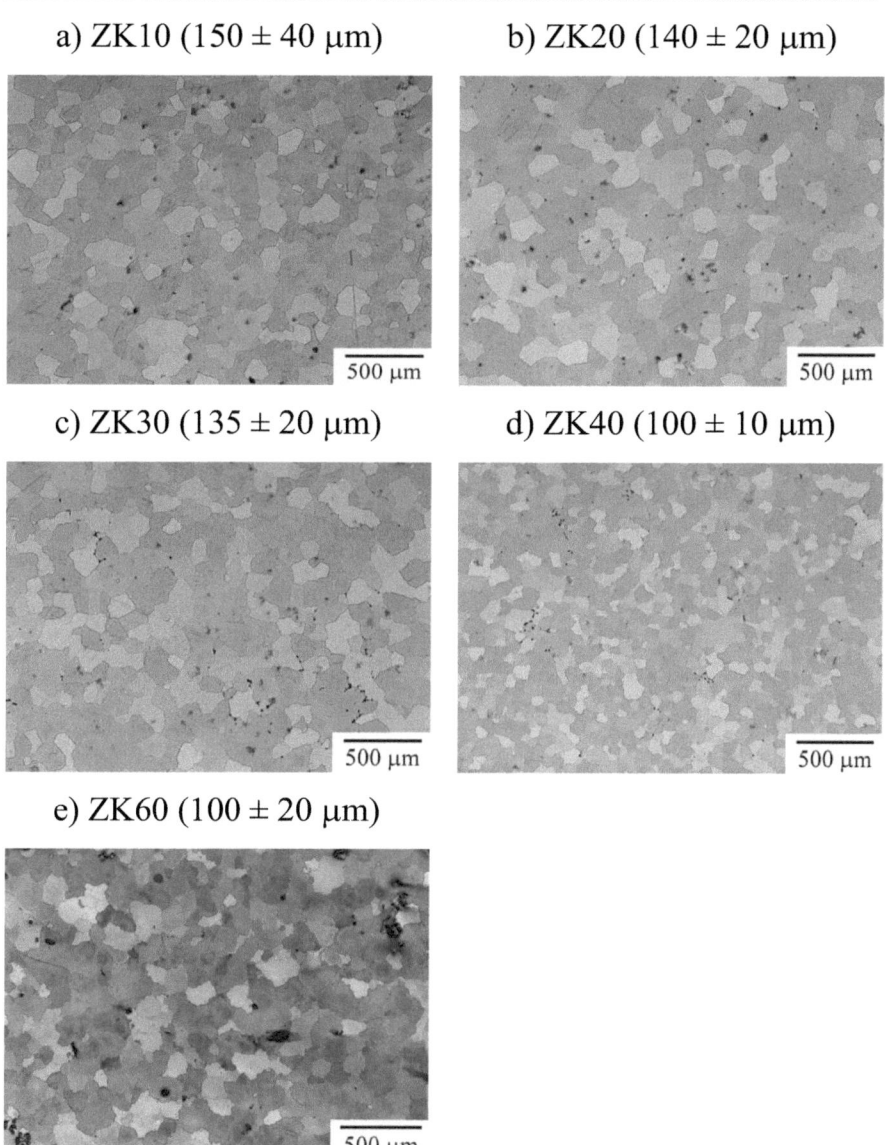

Fig. 4.3 Microstructures of the homogenised ZK-series alloys.

4. Results

Microstructures of the ZEK-series alloys

The addition of Ce-MM to the Mg-Zn-Zr alloys improves the grain refining effect to some extent, see Fig. 4.4 (a)-(c). However, no significant variations with increasing Zn content are found. The microstructures of the ZEK-series alloys also consist of equiaxed, globular grains which are distributed very homogeneously.

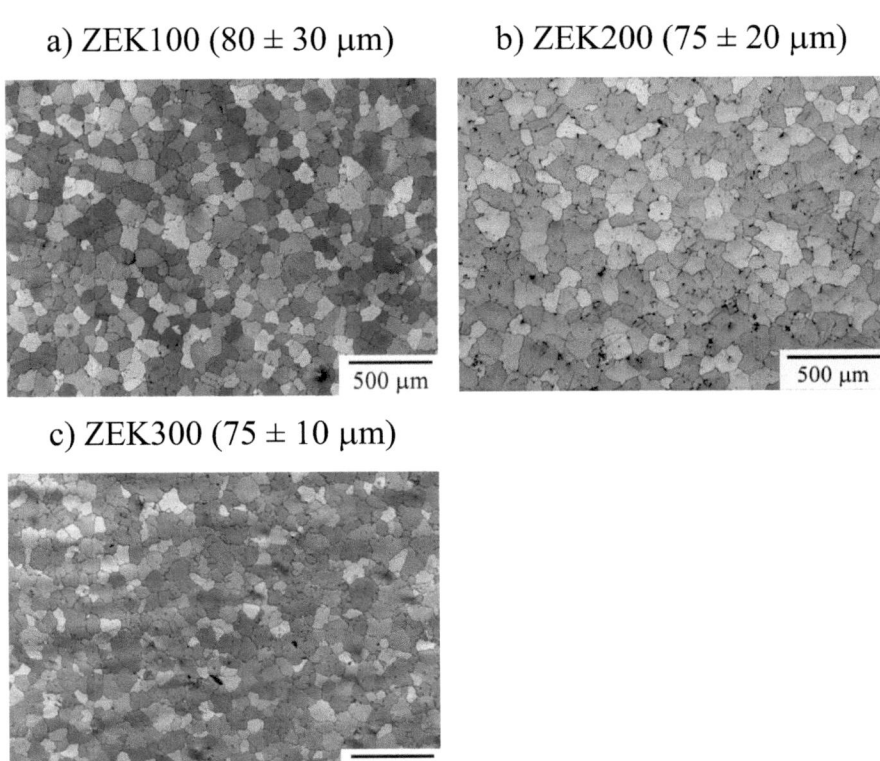

Fig. 4.4 Microstructures of the homogenised ZK-series alloys.

4. Results

For a better appreciation of the effect of the various alloying element additions on the average grain size of all the homogenised alloys, see Fig. 4.5. It is clear that each alloying element addition produces grain refinement of the microstructures to different levels with Zr being the most effective. It needs noting that the (Y) axis of this graph has been modified in the middle for convenience.

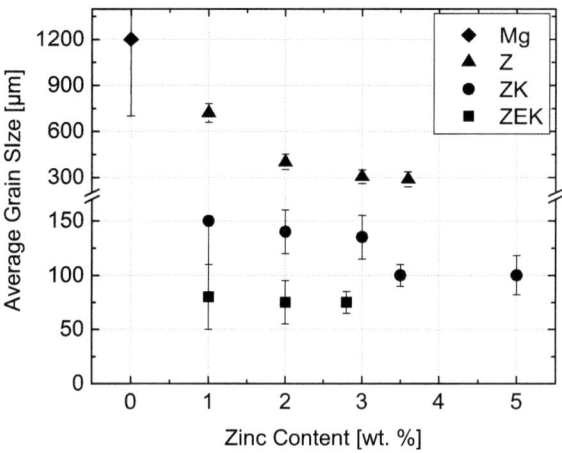

Fig. 4.5. Effect of alloy element additions on the grain sizes of the homogenised alloys.

Phase characterisation of the Z-series alloys

Scanning electron micrographs of the Z-series alloys revealed only some occasional, randomly distributed impurity particles and pores, see Fig. 4.6 for an example.

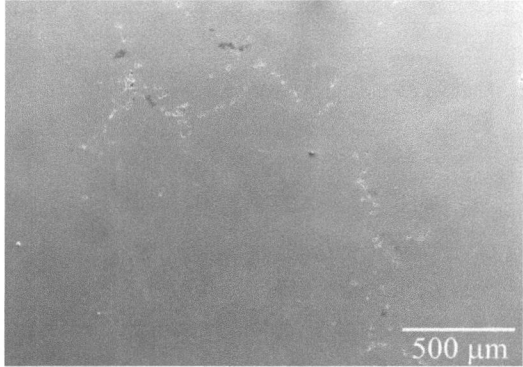

Fig. 4.6. SEM micrograph of the Z4 alloy after the homogenisation treatment.

4. Results

Phase characterisation of the ZK-series alloys

Similar to the Z-series alloys, SEM micrographs of the ZK-series showed the presence of only few particles; however with increasing Zn content the number of particles is slightly increased. Thus only the ZK40 and ZK60 alloys showed a high concentration of particles in this group of alloys, as shown in Figs. 4.7 (a) and (b). The micrographs show the typical precipitates found in these alloys. Most of the particles lie preferentially along the grain boundaries (intergranular), although a few are located inside the grains (intragranular). Some large amorphous particles similar to clusters in very bright contrast were also observed. EDX analysis of these clusters indicated that they correspond to undissolved α-Zr particles, as has frequently been reported for cast Mg-Zn-Zr alloys [52, 106].

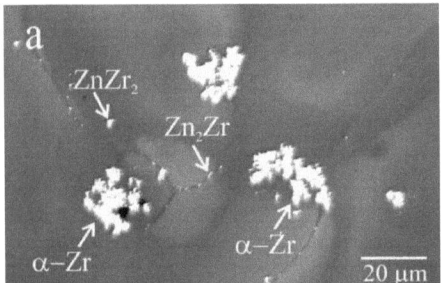 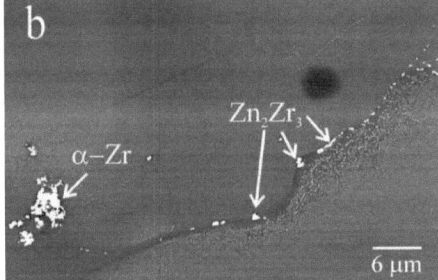

Fig. 4.7. SEM micrographs showing different particles in the microstructures of the alloys ZK40 (a) and ZK60 (b) after the homogenisation treatment

As expected, the Zn-Zr system shows the most stable intermetallic compounds in comparison to the Mg-Zn and Mg-Zr systems. The intergranular and intragranular precipitates found in this series correspond exclusively to this system. EDX analysis revealed that the intergranular particles have a defined composition with high contents of Zr and Zn with the Zr contents being slightly higher than their Zn contents. Considering their stoichiometric composition, it is possible that the intergranular precipitates are the intermetallic phase Zn_2Zr_3 (68 wt. % Zr).

EDX analysis of other intergranular particles reveals that they have higher contents of Zn in comparison with that of Zr. Considering this stoichiometric composition it is very likely that these particles correspond to the intermetallic phase Zn_2Zr (59 wt. % Zn). EDX of the intragranular particles also revealed high contents of Zr and Zn with the Zr contents being higher than that observed for the intergranular particles. Therefore it is very likely that these particles correspond to the intermetallic phase $ZnZr_2$ (74 wt. % Zr).

4. Results

Phase characterisation of the ZEK-series alloys

The microstructures of the ZEK-series alloys consist of α-Mg together with very well outlined grain boundaries; see Figs. 4.4 and 4.8 (a), (b) and (c). This series of alloys show a considerable amount of particles with most located preferentially on the grain boundaries. Moreover they form thick layers of long intergranular particles around the grains, producing the appearance of a network as reported in the literature [35]. The number of particles increased slightly with increasing Zn content.

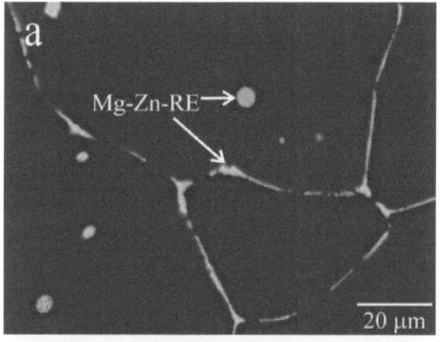
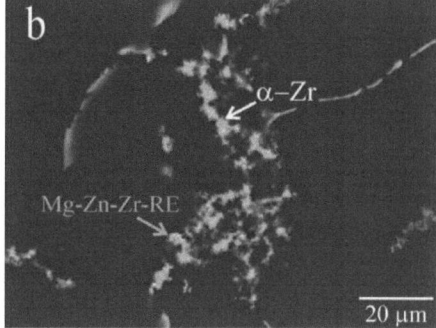
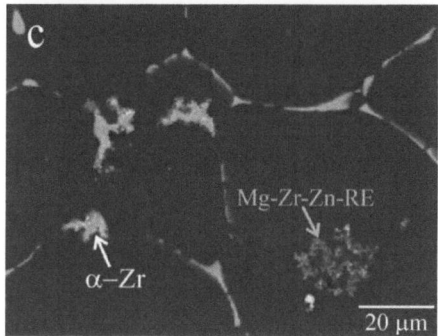

Fig. 4.8. SEM micrographs showing the typical precipitates in the the alloys ZEK100 (a), ZEK200 (b) and ZEK300 (c). Mg-Zn-RE compounds. Compounds rich in Zn (Mg-Zn-Zr-RE). Compounds rich in Zr (Mg-Zr-Zn-RE).

Some brightly imaging particles were found to be undissolved α–Zr but most of the intergranular and intragranular particles are complex compounds of a quaternary system Mg-Zn-Ce-La (called Mg-Zn-RE). EDX analysis of these particles showed a fixed compositional pattern: high contents of Mg and Zn together with about 20 wt. % Ce and La. Other Mg-Zn-RE compounds containing Zr have also been observed in intergranular, intragranular or even with indissoluble α–Zr. Some of them have higher Zn content than that of Zr, called Mg-Zr-Zn-RE compounds (marked in Fig. 4.8). However, others show higher Zr contents than that of Zn called Mg-Zn-Zr-RE compounds (marked in Fig. 4.8).

4. Results

4.2 Extrusion experiments

4.2.1 Extrusion diagrams

For all extrusion trials the measured data have been graphically represented in extrusion diagrams as shown in Fig. 4.9 for the indirectly extruded alloys. During extrusion, the ram speed achieves a stationary stage where the velocity is constant; this is shown in Fig. 3.6 as three plateaus at different levels in a staircase form. The profile temperature registered during extrusion also formed plateaus with constant temperature. The average temperature corresponding to each plateau represents the so-called profile temperature. The same procedure was carried out for the extrusion force and the value corresponding to the average speed was called the steady state force. For a better appreciation of the development of temperature and force as a function of the alloying element addition, the data are described in detail in the next section.

4.2.2 Profile temperatures and steady state forces

The profile temperatures and steady state forces have been plotted against the product speed for the different alloy series as shown in Fig. 4.10. The diagrams show clearly the progressive increase in profile temperature with increased speed. Moreover, the alloying element additions also seem to influence the profile temperatures. Fig. 4.10 (a) shows a significant influence of the Zn addition on the profile temperature, where the highest profile temperature was achieved for the Z4 alloy. Unfortunately, due to a malfunction with the thermocouple, it was not possible to record data for the Z3 alloy. Nevertheless, comparing with pure magnesium, the profile temperature tends to increase clearly with Zn additions.

With the addition of Zr, higher profile temperatures were achieved, see Fig. 4.10 (b). Still, they showed a negligible effect of the Zn addition on the profile temperature. Only the alloys extruded at 10 [m/min] show a slight increase in the temperature with the Zn addition.

With the addition of RE to the ZK-series, the profile temperatures were slightly higher than for the other groups and they correspond to the highest temperatures registered in the present work, see Fig. 4.10 (c). Similar to the ZK-series, the Zn additions did not show any particular influence on the profile temperature.

4. Results

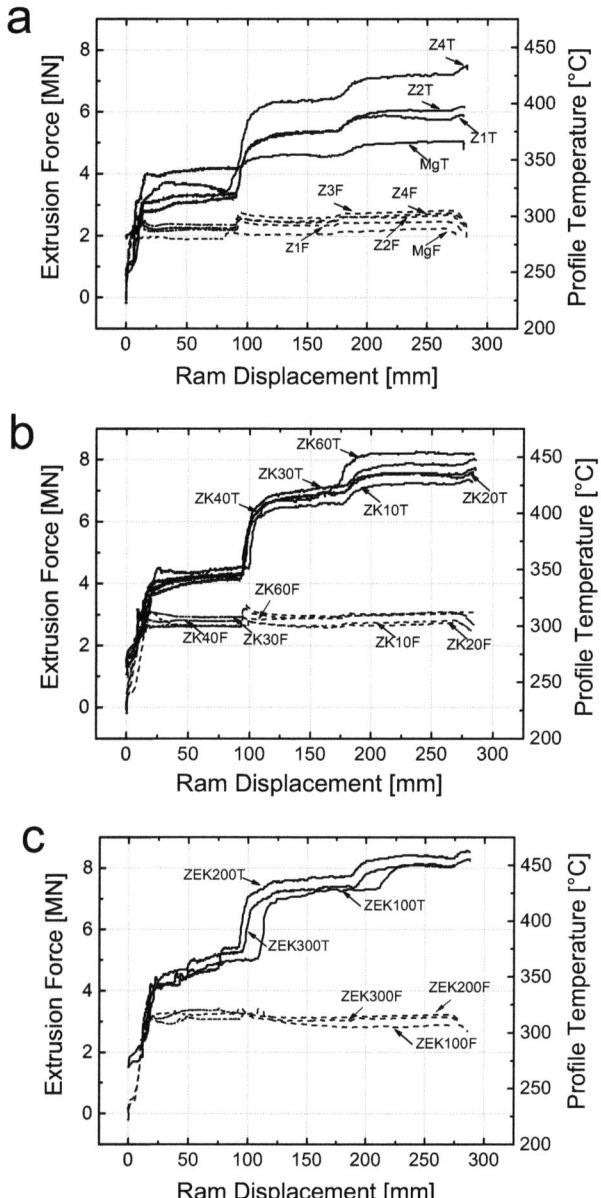

Fig. 4.9. Extrusion diagrams for the indirectly extruded alloys: a) Mg and Z-series, b) ZK-series and c) ZEK-series. T: Profile temperature, F: Extrusion force = Die extrusion force F_D.

4. Results

Concerning the steady state forces, they did not undergo such significant changes as those observed in the profile temperatures. However, some changes can be described as a function of the alloying element addition as well as the extrusion speed. For example, with an increase in the Zn content in the binary alloys the force increased somewhat, although the change is not significant, see 4.10 (a). For these binary alloys a slight increase in the force is also observed with an increase of speed.

On other hand, the addition of Zr to the binary Mg-Zn alloys produces an increase in the force; however, the progressive Zn addition and the increase in speed did not show any separate effect, see Fig. 4.10 (b).

Finally, with the addition of RE to the ZK-series, extrusion forces increased somewhat, see Fig. 4.10 (c); for these alloys, a decrease in the extrusion forces with an increase in the extrusion speed is noted.

4. Results

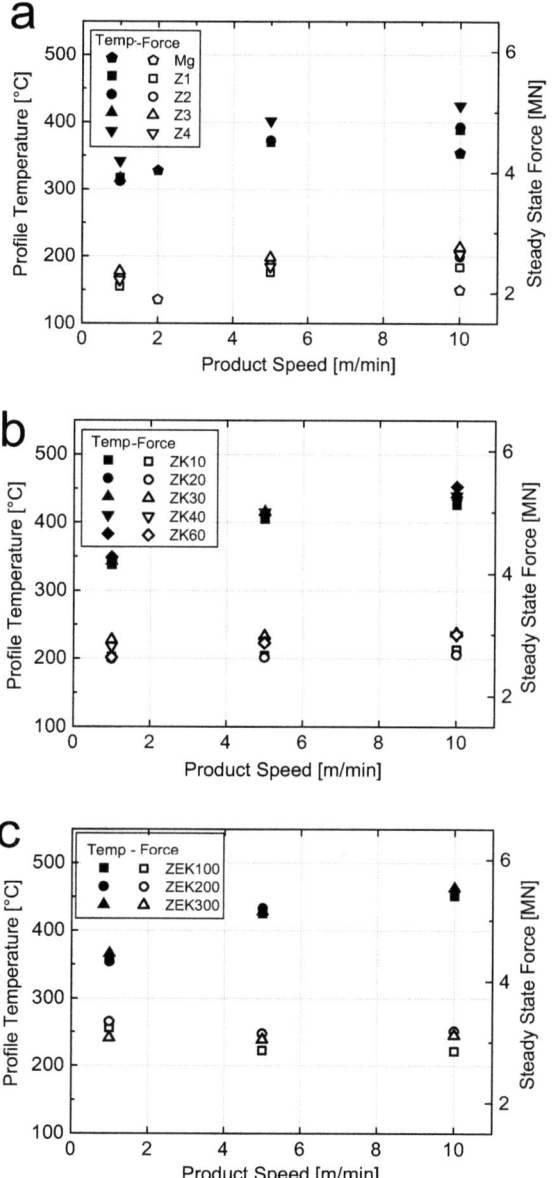

Fig. 4.10. Profile temperature (Temp) and Steady State Force (Force) as a function of extrusion speed for (a) pure Mg and Z-series alloys, (b) ZK-series alloys and (c) ZEK-series alloys.

4. Results

4.3 Microstructures of the extruded profiles

4.3.1 Pure magnesium and the Z-series alloys

Indirect extrusion of pure Mg as well as the Z-series alloys resulted in very homogeneous, i.e. well recrystallised and grain refined microstructures as shown in Figs. 4.11, 4.12 and 4.13. The indirectly extruded Z-series alloys show in general a grain coarsening effect at higher profile speeds, whereas the grain size of pure Mg was negligibly affected by the extrusion speed.

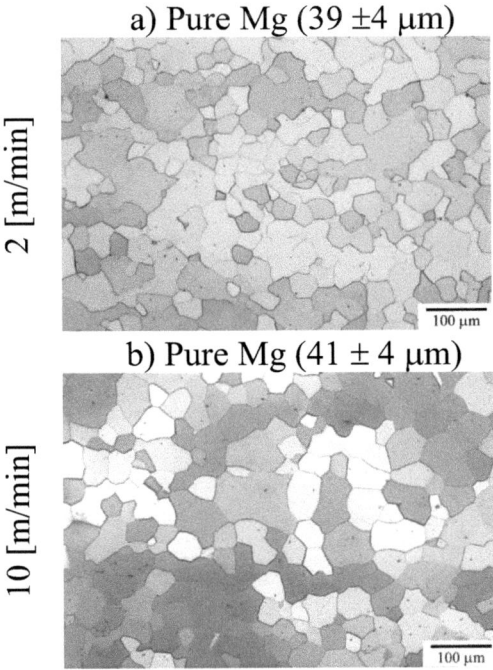

Fig. 4.11. Microstructure and grain size (extrusion direction →) of indirectly extruded pure magnesium at different profile speeds.

4. Results

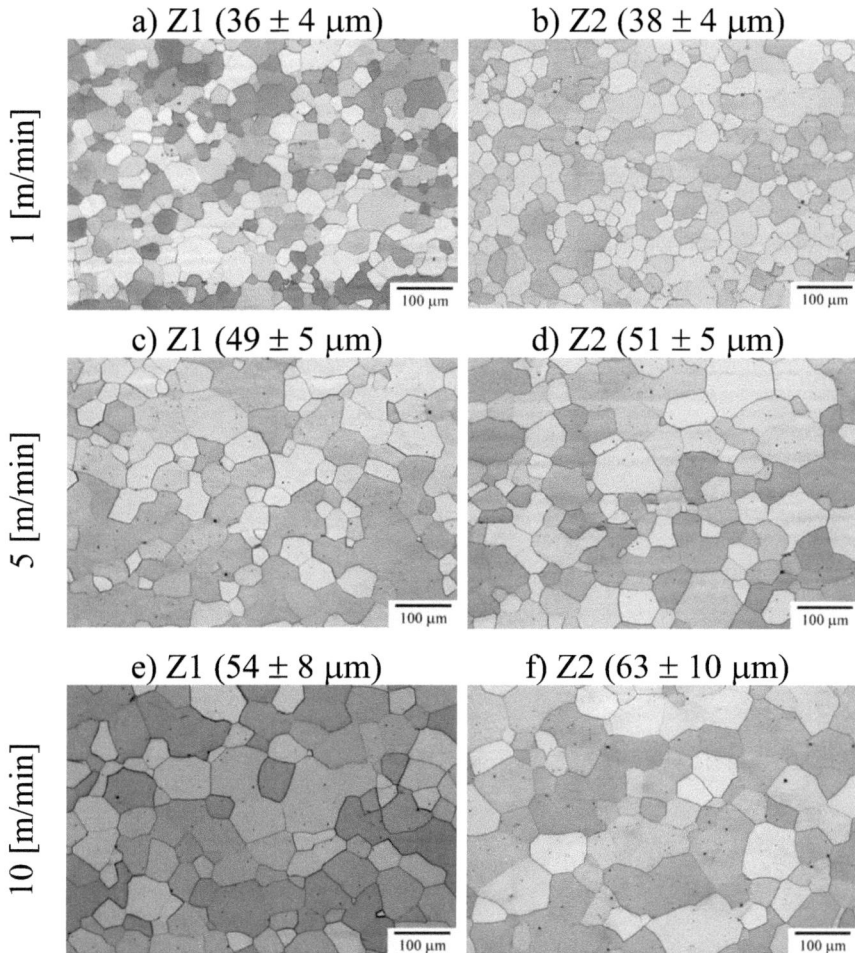

Fig. 4.12. Microstructure and grain size (extrusion direction →) of the indirectly extruded Z1 and Z2 alloys at different profile speeds: (1, 5 and 10 [m/min]).

4. Results

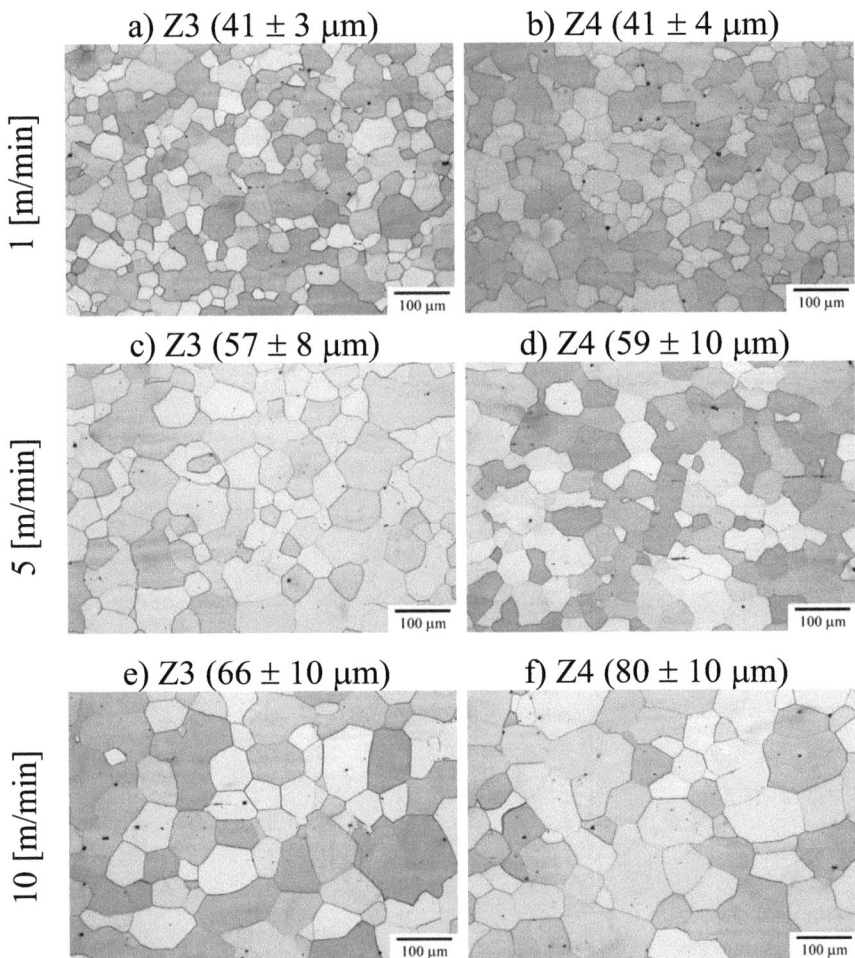

Fig. 4.13. Microstructure and grain size (extrusion direction →) of the indirectly extruded Z3 and Z4 alloys at different profile speeds: (1, 5 and 10 [m/min]).

The measured grain sizes of the indirectly extruded pure Mg and Z-series alloys are summarised in Fig. 4.14. Alloys extruded at 10 [m/min] show clearly a grain coarsening effect with the Zn addition. This effect is even more contrasted in comparison with grain size of pure Mg.

4. Results

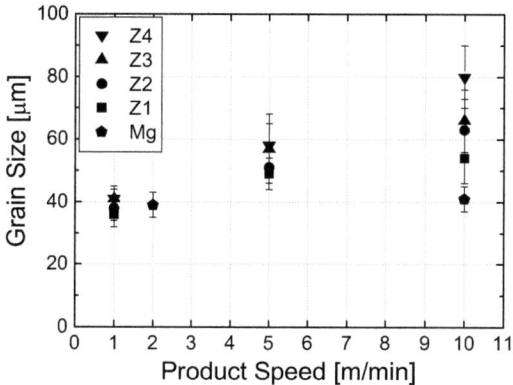

Fig. 4.14. Grain size as a function of the product speed for the indirectly extruded pure Mg and Z-series alloys.

Phase characterisation of the extruded Z-series alloys

The microstructures of the extruded Z-series contained only some occasional, randomly distributed impurity particles. Their presence was only slightly increased with a high content of Zn, thus the extruded Z1, Z2 and Z3 alloys are almost free of particles and only the extruded Z4 alloy showed some of them, see Fig. 4.15. Some of these particles are arranged in bands along the extrusion direction and other fine particles are located along the grain boundaries of the newly recrystallised grains. EDX analysis of these particles indicates that they correspond to the intermetallic phase MgZn (74 wt.% Zn). Particles found in the other Z-series alloys also correspond to the same phase.

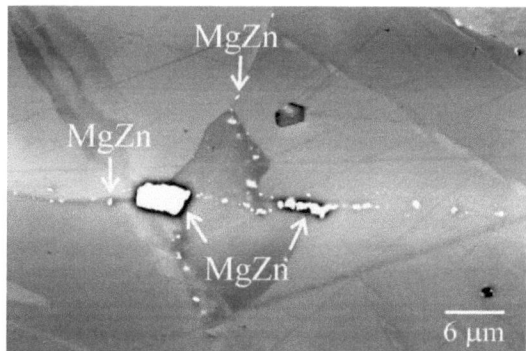

Fig. 4.15. SEM micrograph of particles in the indirectly extruded Z4 alloy at 10 [m/min].

4. Results

4.3.2 ZK-series alloys

Figs. 4.16, 4.17 and 4.18 show the microstructures of the indirectly extruded ZK-series alloys. The microstructures of this series, without exception, are inhomogeneous, i.e. recrystallisation has not been completed. Average grain size values are based on the recrystallised grains.

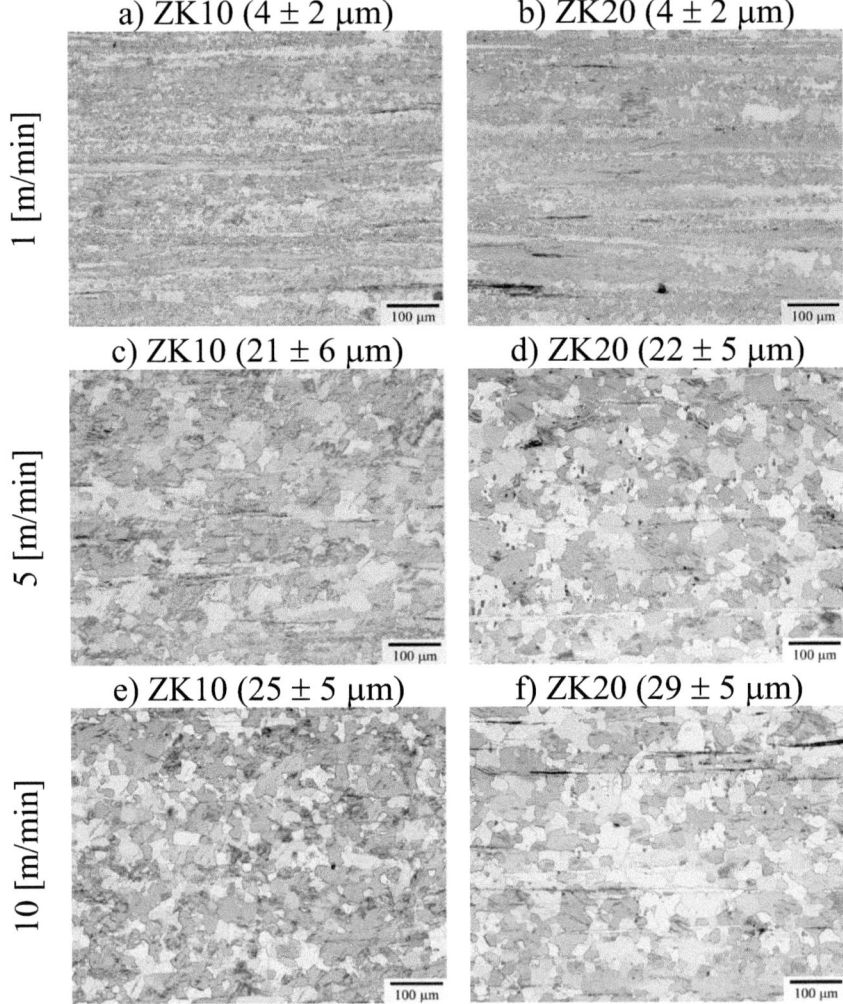

Fig. 4.16. Microstructure and grain size from longitudinal sections (extrusion direction →) of the indirectly extruded ZK10 and ZK20 alloys at different profile speeds: (1, 5 and 10 [m/min]).

4. Results

This incomplete recrystallisation process can be especially observed in the microstructures of the ZK-series alloys extruded at 1 [m/min], in which large unrecrystallised grains are surrounded by new recrystallised grains. Alloys with the highest Zn content also showed large unrecrystallised grains, e.g. ZK60 see Figs. 4.18 (a)-(c).

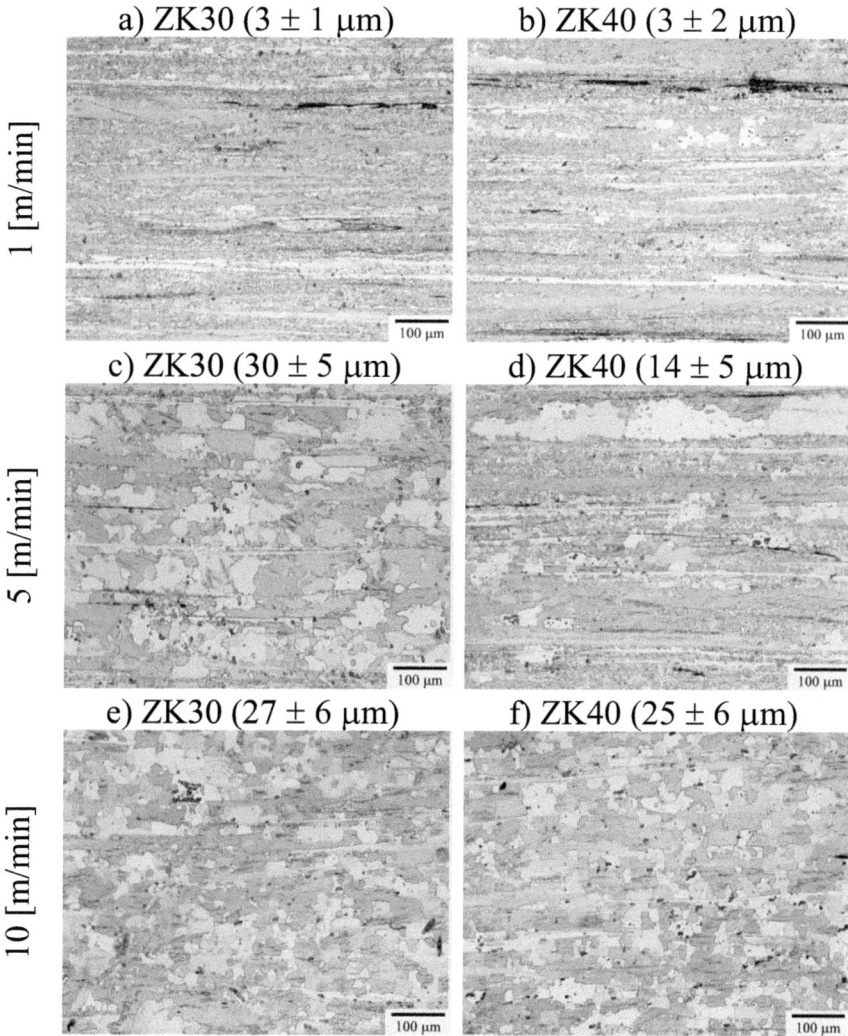

Fig. 4.17. Microstructure and grain size from longitudinal sections (extrusion direction →) of the indirectly extruded ZK30 and ZK40 alloys at different profile speeds: (1, 5 and 10 [m/min]).

69

4. Results

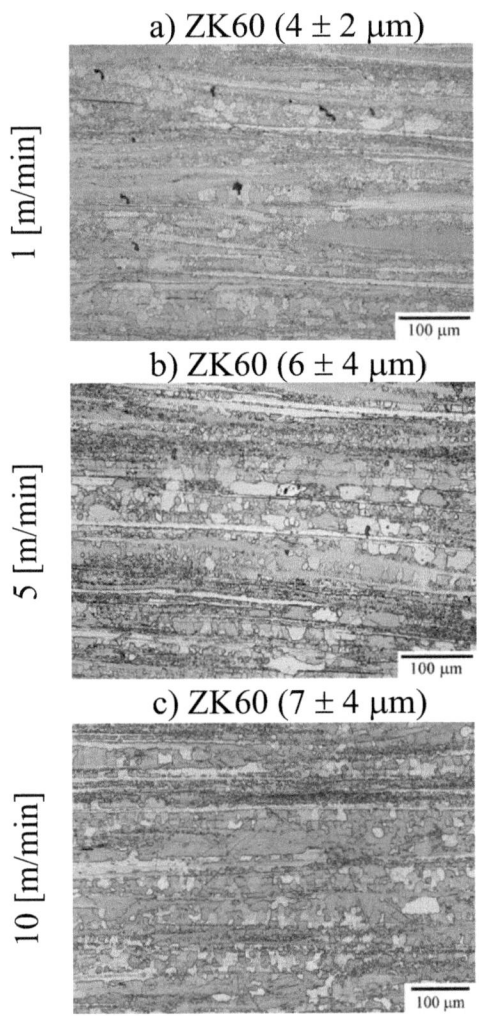

Fig. 4.18. Microstructure and grain size from longitudinal sections (extrusion direction →) of indirectly extruded ZK60 alloy at different profile speeds.

Fig. 4.19 summarises the measured grain sizes of the indirectly extruded ZK-series alloys. Grain sizes were coarser for the alloys extruded at higher speed. However, there is no ongoing increase in the grain size with the Zn addition as observed in the Z-series. It is interesting to note that in contrast to the Z-series, the Zn addition in the ZK-series shows a grain growth inhibition effect, thus ZK60 alloys show the finest grain size of this series. This finding corresponds with the addition of Zr as alloying element in comparison to the Z -series.

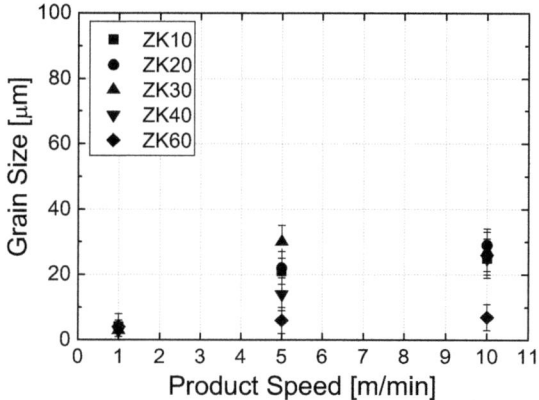

Fig. 4.19. Grain size as a function of the product speed for the indirectly extruded ZK-series alloys.

Phase characterisation of extruded ZK-series

Similar to cast ZK-series alloys, SEM observation of the ZK-series alloys extruded at different speeds reveals that the greatest number of particles is observed in the alloys with the highest Zn content, i.e. ZK40 and ZK60 alloys, see Fig. 4.20 (b) – (d). The rest showed only sparsely distributed particles arranged in bands along the extrusion direction, as is shown in Figs. 4.16 and 4.17. EDX analysis reveals that they are made up of undissolved α-Zr particles, as shown in Fig. 4.20 (a) for the ZK10 alloy. These α-Zr particles were already present in the as cast condition.

These bands of particles were also observed in the ZK40 and ZK60 alloys extruded at different speeds. However, EDX indicates that they consist of undissolved α-Zr particles and precipitates of the intermetallic phases Zn_2Zr_3, $ZnZr_2$ and Zn_2Zr.

The principal difference with respect to the cast condition is a massive presence of fine precipitates developed in the newly recrystallised microstructure; therefore, it can be assumed that their appearance was definitely influenced by the process. Most of these new precipitates correspond to the intermetallic phase Zn_2Zr.

4. Results

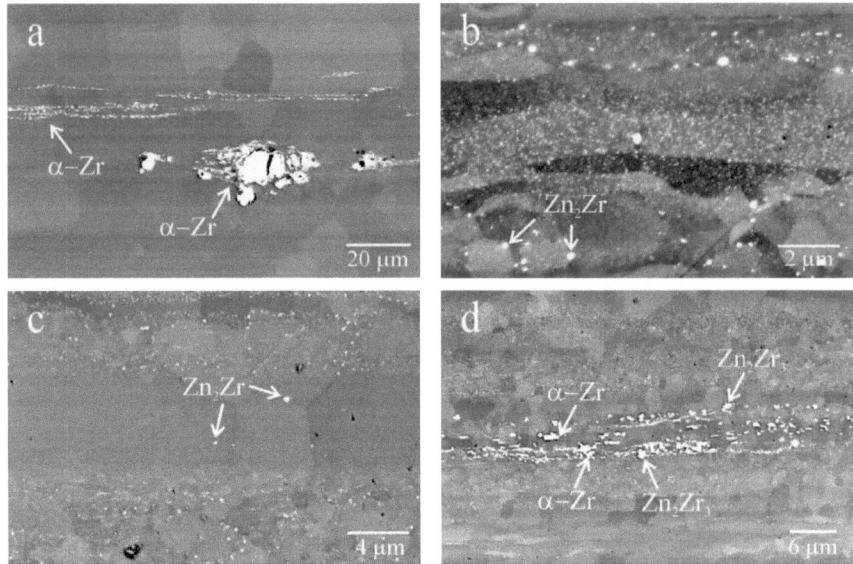

Fig. 4.20. SEM micrographs of the alloys (a) ZK10, (b) ZK40, (c) and (d) ZK60, indirectly extruded at 10 [m/min]:

4. Results

4.3.3 ZEK-series alloys

Figs. 4.21 and 4.22 show the microstructures of the indirectly extruded ZEK-series alloys. In comparison to the ZK-series alloys, these alloys only contain an additional amount of RE. The resulting microstructures are more homogeneous than their counterparts without RE, however a large fraction of large unrecrystallised grains is also observed, especially in the alloys extruded at 1 [m/min].

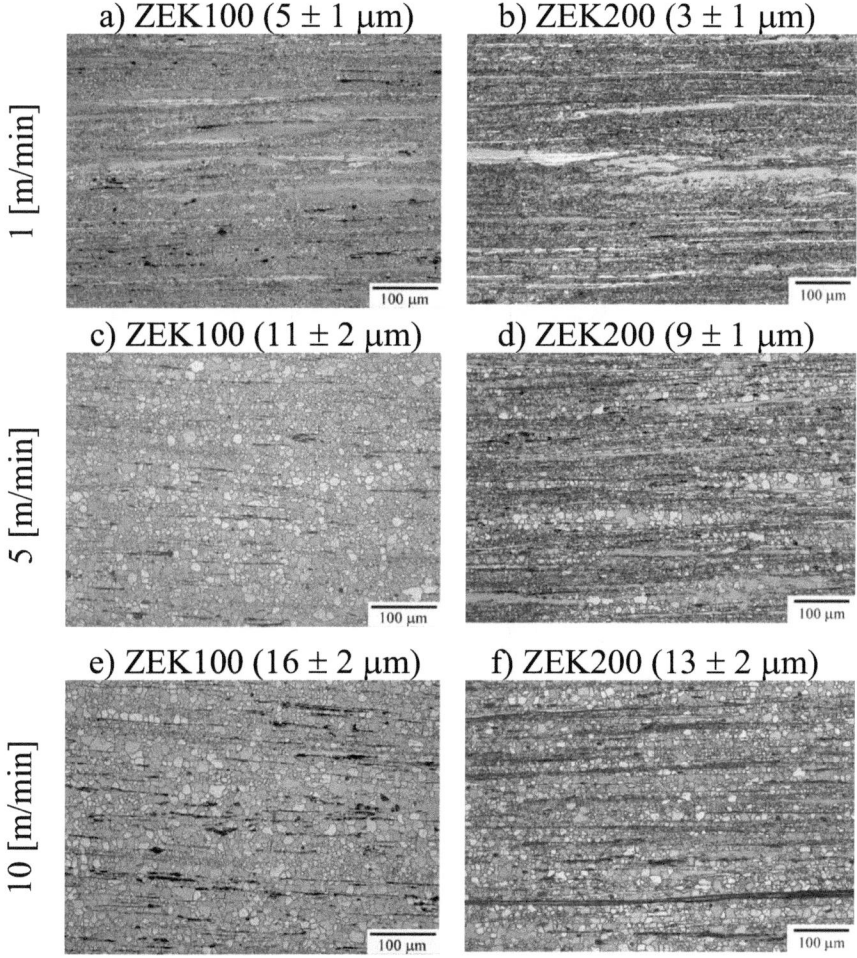

Fig. 4.21. Microstructure and grain size from longitudinal sections (extrusion direction →) of the indirectly extruded ZEK100 and ZEK200 alloys at different profile speeds: (1, 5 and 10 [m/min]).

4. Results

With increase in the extrusion speed, there is also a slight grain coarsening; however in comparison to the other series, grain coarsening is strongly inhibited. This is remarkable, because these alloys registered also the highest profile temperatures.

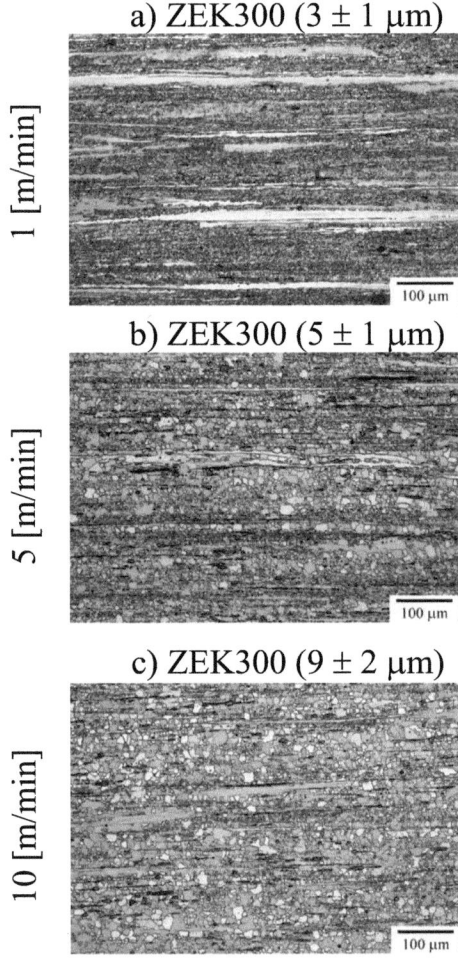

Fig. 4.22. Microstructure and grain size from longitudinal sections (extrusion direction →) of indirectly extruded ZEK300 alloy at different profile speeds.

4. Results

On other hand, Zn additions did not show any considerable influence on the grain size. This is better viewed in Fig. 4.23, where the resulting grain sizes for this series are presented. These alloys show the finest grain size among all the alloys studied.

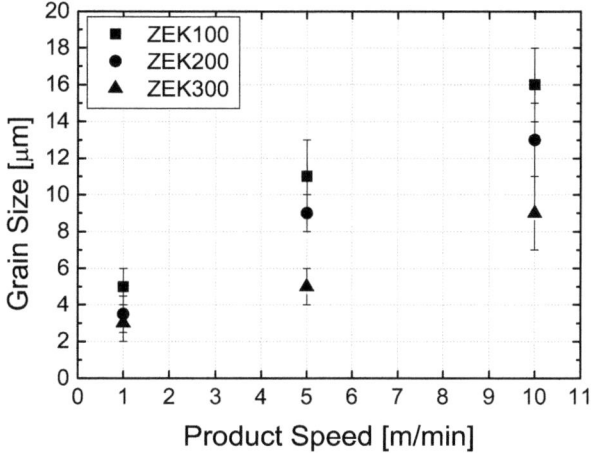

Fig. 4.23. Grain size as a function of extrusion speed for the ZK-series alloys.

Phase characterisation of the extruded ZK-series alloys

In comparison with the ZK-series a higher fraction of bands of precipitates is visible along the extrusion direction; see Figs. 4.21, 4.22 and 4.24. They are present in all the ZEK-series alloys for all the extrusion conditions. SEM observations and EDX analysis indicated that these bands contain the particles observed in the as cast condition, i.e. Mg-Zn-RE, Mg-Zr-Zn-RE and Mg-Zn-Zr-RE compounds, as well as undissolved α-Zr particles. It is assumed that the coarse particles observed in the as cast condition were broken and dispersed in the matrix uniformly during the extrusion process, as reported previously [56, 107]. In contrast to the other series, the ZEK-series alloys showed no visible increase of particles with increasing extrusion speed or with the progressive Zn additions.

4. Results

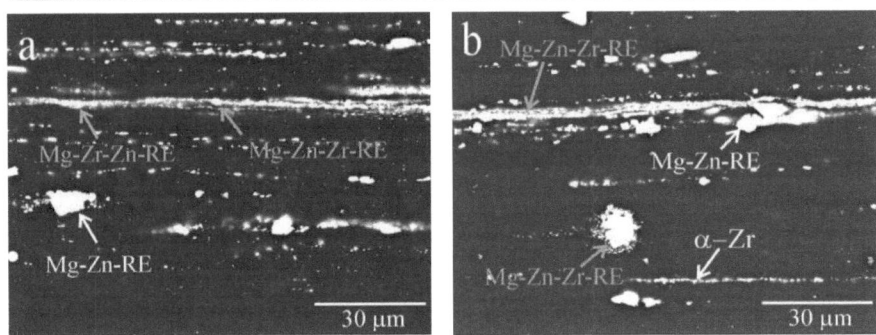

Fig. 4.24. SEM micrographs showing particles in ZEK100 (a) and ZEK300 (b) after indirect extrusion at 10 [m/min]. Compounds rich in Zn (Mg-Zn-Zr-RE) and compounds rich in Zr (Mg-Zr-Zn-RE).

4. Results

4.4 Textures of extruded Mg-Zn based alloys

In the following, the results of texture measurements on the extruded profiles will be presented. This analysis has been focussed on the materials represent the influence of the lowest and the highest extrusion speed as well as the minimum and the maximum temperature achieved during the process, i.e. alloys extruded at 1 and 10 [m/min]. These conditions represent also the principal microstructural changes undergone during extrusion. Each alloy group will be described separately and at the end of this section the principal changes will be summarised in a table.

The intensity levels are represented by a grey scale and the darkest one indicates the maximum intensity of a preferred crystallographic plane, which is oriented tangentially at any point on the surface perpendicular to the extrusion direction. The maximum intensity value is also indicated on the left of the inverse pole figure in arbitrary units of multiples of random distribution [m.r.d]. The three corners of the triangle represent the components of the basal plane (0001) and the principal prismatic planes, i.e. the $(10\bar{1}0)$ plane and the $(11\bar{2}0)$ plane). If a maximum intensity is shown between these components, then this indicates that the principal component correspond to another crystallographic plane. Some possibilities are marked as points in Fig. 2.19. On other hand it is also possible to observe a scatter of one component towards another.

4.4.1 Pure Mg and the Z-series alloys

Fig. 4.25 shows the inverse pole figures for pure Mg and the Z-series alloys indirectly extruded at speeds of 1 and 10 [m/min], respectively (with a slight difference in the case of pure Mg extruded at 2 [m/min]).

Pure Mg

The textures of pure Mg in both extruded conditions are very similar; see Fig. 4.25 (a) and (b). Though the continuously distributed pole densities between the $\langle 10\bar{1}0 \rangle$ and the $\langle 11\bar{2}0 \rangle$ prismatic poles are found in the inverse pole figure in the extrusion direction, the maximum intensities are preferentially oriented towards the $\langle 11\bar{2}0 \rangle$ poles. In the case of pure Mg extruded at 10 [m/min], the highest intensity is not located exactly at the $\langle 11\bar{2}0 \rangle$ pole, but tilted ~ 15° from this pole. Pure Mg extruded at 2 [m/min] also shows this broadening but with a lower tilt angle. However, both samples show a broad intensity distribution tilted ~ 15° from the $\langle 10\bar{1}0 \rangle$ prismatic pole.

4. Results

Z-series alloys

The highest intensities of the extruded Z-series alloys are also found distributed continuously between the $\langle 10\bar{1}0 \rangle$ and the $\langle 11\bar{2}0 \rangle$ prismatic poles; see Fig. 4.25 (c)-(j). Similar to pure Mg, the Z-series alloys extruded at 1 [m/min] also show high intensity peaks tilted ~15° from the $\langle 11\bar{2}0 \rangle$ prismatic pole. The component tilted from the $\langle 10\bar{1}0 \rangle$ prismatic pole is also found, but for these alloys it is very weak and tends to vanish completely with an increase in the Zn content, see Fig. 4.25 (c),(e), (g) and (i).

In contrast, the Z-series alloys extruded at 10 [m/min] show a perfectly outlined $\langle 11\bar{2}0 \rangle$ component, see Fig. 4.25 (d), (f), (h) and (j).

Independently of the extrusion speed, the Z-series alloys tend to extend the distribution of the intensities from the prismatic poles $\langle 10\bar{1}0 \rangle$ and $\langle 11\bar{2}0 \rangle$ towards the basal component $\langle 0001 \rangle$, as the content of Zn increases.

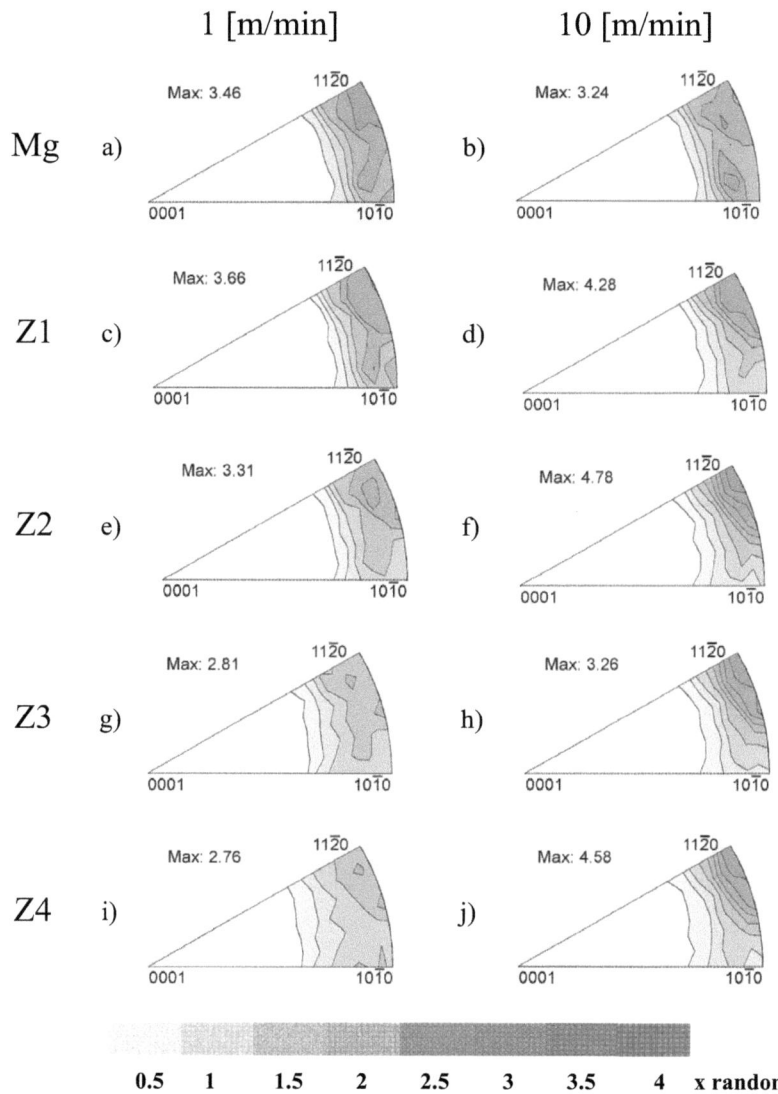

Fig. 4.25. Inverse pole figures of pure Mg indirectly extruded at 2 (a) and 10 (b) [m/min] and Z-series alloys indirectly extruded at 1 and 10 [m/min] (c)-(j).

4.4.2 ZK-series alloys

The inverse pole figures of the ZK-series alloys indirectly extruded at 1 and 10 [m/min] are shown in Fig. 4.26. They show higher intensities than their counterparts without Zr. Profiles extruded at 1 [m/min] show considerably higher intensities (between 12.3-17.6 [m.r.d], see Fig. 4.26 (a),(c),(e), (g) and (i)) in comparison with those extruded at 10 [m/min] (between 5.1-16.1 [m.r.d], see Fig. 4.26 (b), (d), (f), (h) and (j)). All samples, without exception, show the highest intensity focussed on the $\langle 10\bar{1}0 \rangle$ pole, especially for the alloys extruded at 1 [m/min]. However, it is also possible to observe the development of a double texture component distributed along the curvature between the poles $\langle 10\bar{1}0 \rangle$ and $\langle 11\bar{2}0 \rangle$ in the alloys with low Zn content extruded at 10 [m/min].

If the effect of the progressive Zn addition is separately observed in Fig. 4.26, it is clear that the Zn addition promotes the development of a strong $\langle 10\bar{1}0 \rangle$ fibre-texture component, and a double texture component is inhibited. Concerning the effect of the Zn addition on the intensities, it can be seen that whereas alloys extruded at 1 [m/min] do not show any considerable change, alloys extruded at 10 [m/min] show a significant increase in the intensities with Zn additions.

In addition, a component located 25° from the $\langle 11\bar{2}0 \rangle$ pole towards the basal direction can be observed. This component corresponds to the $\langle 11\bar{2}1 \rangle$ pole; its position is indicated with black arrows in Fig. 4.26. It can be observed that Zn additions increase its intensity independently of the extrusion speed; thus the presence of this component is better outlined by ZK40 and ZK60 alloys, see Fig. 4.26 (g), (i) and (j).

4. Results

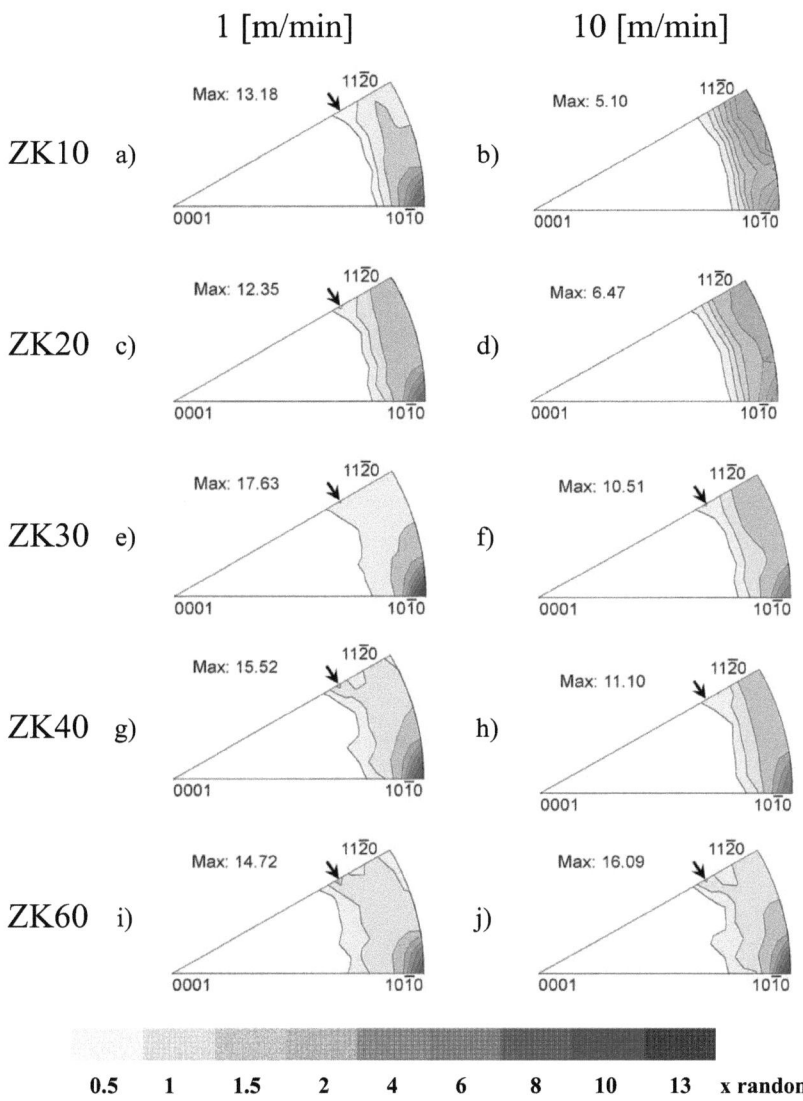

Fig. 4.26. Inverse pole figures of the ZK-series alloys indirectly extruded at 1 [m/min] and 10 [m/min]. The presence of a texture component on the pole $\langle 11\bar{2}1 \rangle$ is marked with black arrows.

4.4.3 ZEK-series alloys

The inverse pole figures for the ZEK-series alloys indirectly extruded at 1 and 10 [m/min] are shown in Fig. 4.27. They registered the highest intensities of all the alloys in this work.

This group of alloys shows similar characteristics to those of the ZK-series alloys. The profiles extruded at 1 [m/min] show considerably higher intensities (between 17.7-23.6 [m.r.d]) in comparison to the profiles extruded at 10 [m/min] (between 6.3-10 [m.r.d]). These gradients in intensities are very high and correspond to the highest ones found in this work. Again, the highest intensities are focussed on the $\langle 10\bar{1}0 \rangle$ pole especially for the alloys extruded at 1 [m/min], whereas the weaker intensities observed for the alloys extruded at 10 [m/min] are accompanied by a broadening of the intensities toward the $\langle 11\bar{2}0 \rangle$ pole. However, the intensity levels of this broadening are lower in comparison to those observed in the ZK-series alloys, for which a very well outlined double fibre-texture was observed. Actually, this broadening of the intensity distribution tends to orient more towards the $\langle 11\bar{2}1 \rangle$ pole (see black arrows in Fig. 4.27) than towards the $\langle 11\bar{2}0 \rangle$ pole, an effect similar to that observed in the ZK-series alloys.

The textures of the ZEK-series alloys do not show any change correlated with the progressive addition of Zn. Still, the three different alloys show particular characteristics related with their intensities independently of extrusion speed, whereas ZEK100 alloys showed the lowest intensities of this group, ZEK200 alloys show a pronounced increase of ~ 4 [m.r.d], followed by a slight fall of the intensities of the ZEK300 alloys.

4. Results

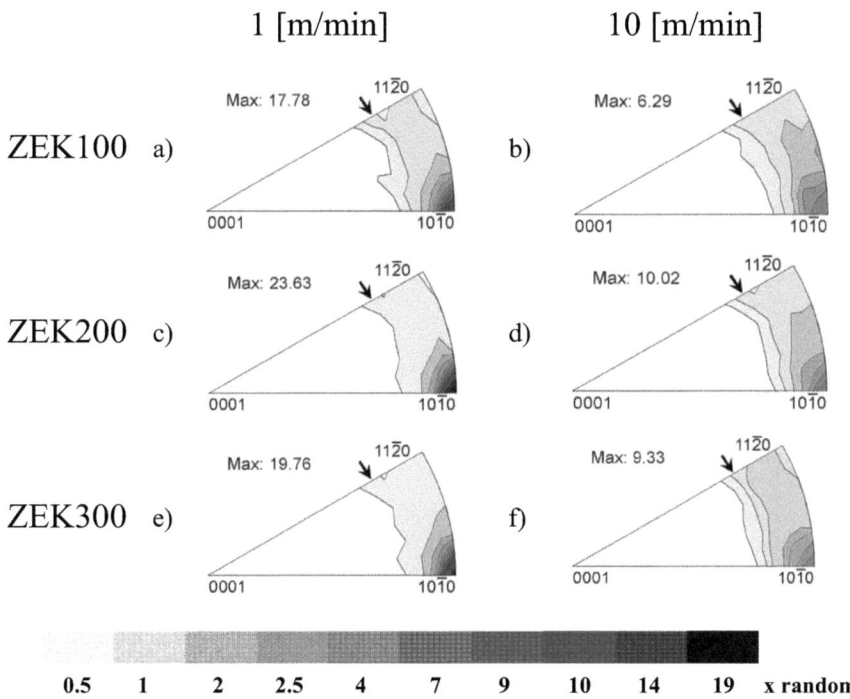

Fig. 4.27. Inverse pole figures of the ZEK-series alloys indirectly extruded at 1 [m/min] and 10 [m/min]. The $\langle 11\bar{2}1 \rangle$ pole is marked with a black arrow.

4. Results

Table 4.1 shows a summary of the principal changes observed for the different alloys. The main texture components with their corresponding intensities are highlighted.

Table 4.1: Summary of the principal texture components and maximum intensities of the extruded alloys. (*) extruded at 2 [m/min].

Extrusion speed [m/min]	Alloy	Texture components	Maximum intensities [m.r.d]
1	Mg*	$\langle 10\bar{1}0 \rangle + \langle 11\bar{2}0 \rangle$ (both tilted ~15°)	3.46
	Z1	$\langle 10\bar{1}0 \rangle + \langle 11\bar{2}0 \rangle$ (both slightly tilted)	3.66
	Z2	$\langle 11\bar{2}0 \rangle$ (tilted)	3.31
	Z3	$\langle 11\bar{2}0 \rangle$ (tilted)	2.81
	Z4	$\langle 11\bar{2}0 \rangle$ (tilted)	2.76
	ZK10	$\langle 10\bar{1}0 \rangle$	13.18
	ZK20	$\langle 10\bar{1}0 \rangle$ + weak $\langle 11\bar{2}0 \rangle$ & $\langle 11\bar{2}1 \rangle$	12.35
	ZK30	$\langle 10\bar{1}0 \rangle$ + weak $\langle 11\bar{2}1 \rangle$	17.63
	ZK40	$\langle 10\bar{1}0 \rangle + \langle 11\bar{2}1 \rangle$	15.52
	ZK60	$\langle 10\bar{1}0 \rangle + \langle 11\bar{2}1 \rangle$	14.72
	ZEK100	$\langle 10\bar{1}0 \rangle$ + weak $\langle 11\bar{2}1 \rangle$	17.78
	ZEK200	$\langle 10\bar{1}0 \rangle$ + weak $\langle 11\bar{2}1 \rangle$	23.63
	ZEK300	$\langle 10\bar{1}0 \rangle$ + weak $\langle 11\bar{2}1 \rangle$	19.76
10	Mg	$\langle 10\bar{1}0 \rangle + \langle 11\bar{2}0 \rangle$ (both tilted ~15°)	3.24
	Z1	$\langle 11\bar{2}0 \rangle$	4.28
	Z2	$\langle 11\bar{2}0 \rangle$	4.78
	Z3	$\langle 11\bar{2}0 \rangle$	3.26
	Z4	$\langle 11\bar{2}0 \rangle$	4.58
	ZK10	$\langle 10\bar{1}0 \rangle + \langle 11\bar{2}0 \rangle$	5.10
	ZK20	$\langle 10\bar{1}0 \rangle + \langle 11\bar{2}0 \rangle$	6.47
	ZK30	$\langle 10\bar{1}0 \rangle + \langle 11\bar{2}0 \rangle$	10.51
	ZK40	$\langle 10\bar{1}0 \rangle + \langle 11\bar{2}0 \rangle$	11.10
	ZK60	$\langle 10\bar{1}0 \rangle$ + weak $\langle 11\bar{2}1 \rangle$	16.09
	ZEK100	$\langle 10\bar{1}0 \rangle$	6.29
	ZEK200	$\langle 10\bar{1}0 \rangle$ + weak $\langle 11\bar{2}1 \rangle$	10.02
	ZEK300	$\langle 10\bar{1}0 \rangle$ + weak $\langle 11\bar{2}1 \rangle$	9.33

4.5 Mechanical properties of the extruded alloys

4.5.1 Pure magnesium and the Z-series alloys

Representative stress-strain curves in tension and compression tests of pure Mg and the Z-series alloys indirectly extruded at different speeds are shown in Fig. 4.28 (a)-(c). In general, all the alloys in all conditions show remarkable differences between the tensile yield strength (TYS) and the compressive yield strength (CYS).

The tensile flow curves of this group of alloys for all extrusion conditions exhibit parabolic hardening typical of slip-dominated deformation, whereas compressive stress-strain curves show the sigmoidal (S-shaped) hardening associated with a twinning-dominated mechanism, in agreement with the texture. In contrast to the tensile flow curves, compressive flow curves are characterised by a slight sharp yield point followed by a decrease in the stress where a plastic strain responds at an almost constant stress.

For some tensile flow curves, micro-yielding occurred in the early stages of plastic deformation; see detail in Fig. 4.28 (d). This effect is associated with a considerable decrease in the TYS to levels even below the CYS. In contrast to the fall of the TYS, the CYS is constant for all these indirectly extruded alloys. Micro-yielding was shown principally by the indirectly extruded pure magnesium profiles and also by the low Zn-containing Z-series alloys extruded at speeds of 5 and 10 [m/min]. Z-series alloys indirectly extruded at a speed of 1 [m/min] did not show micro-yielding phenomena.

4. Results

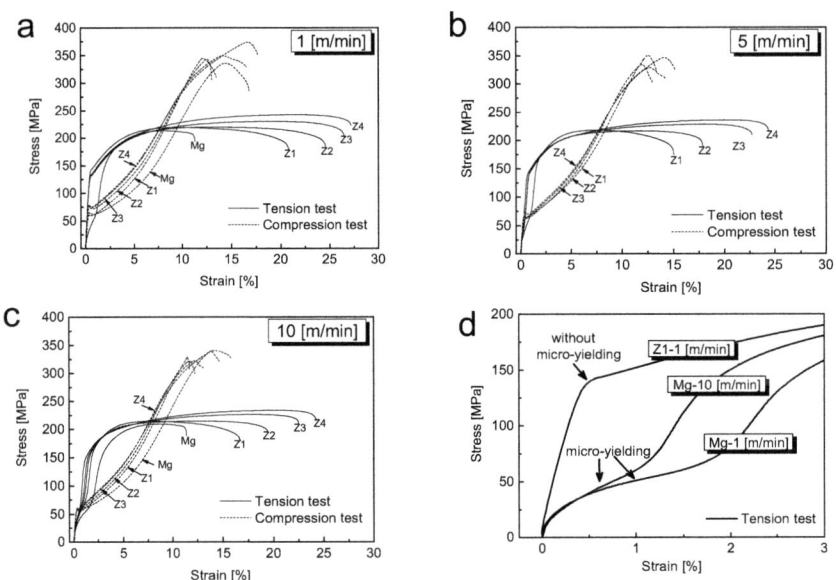

Fig. 4.28. Flow curves in tension and compression for the Z-series alloys indirectly extruded at speeds of 1 (a), 5 (b) and 10 (c) [m/min], and pure Mg indirectly extruded at 2 (a) and 10 (c) [m/min]. Micro-yielding is represented by pure Mg flow curves in (d).

For a better visualisation of the tensile and compressive yield strengths, these values are plotted separately as a function of the extrusion speed; see Figs. 4.29 (a) and (b). All the numerical values are also summarised in Table 4.2. As previously mentioned, pure Mg and the Z-series alloys indirectly extruded at 5 and 10 [m/min] show the micro-yielding phenomenon, which is represented as a drastic drop in the TYS (even below the CYS) in this plot. To avoid confusion, values associated with micro-yielding are not considered in describing the yield strength behaviour; thus the data for the Z-series alloys indirectly extruded at 1 [m/min] are the only ones available for this purpose. The TYS values are higher than that for CYS, thus the yield asymmetry is very marked, see also $\Delta\sigma(\Delta\sigma = TYS - CYS)$ in Table 4.2. However, the progressive increase in Zn content did not show any significant effect on the yield strength.

The UCS values were also higher than that of the UTS, see Fig. 4.29 (c). However, the UCS values tend to reduce slightly at higher extrusion speeds, whereas the UTS did not show any change. Slight increases in the UTS and UCS are observed with increasing Zn content.

4. Results

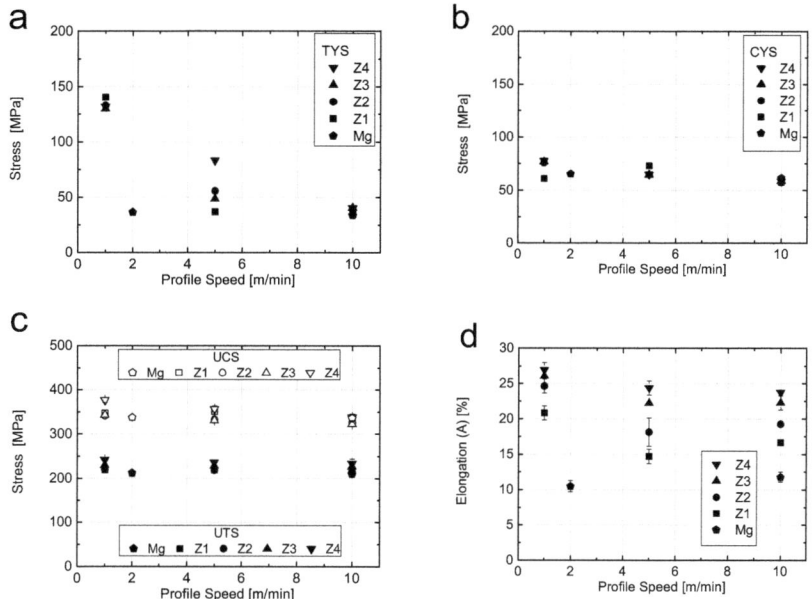

Fig. 4.29. Mechanical properties of pure Mg and the Z-series alloys as a function of extrusion speed: (a) Tensile yield strength TYS, (b) Compressive yield strength CYS, (c) Ultimate tensile strength UTS and Ultimate compressive strength UCS, (d) Elongation A.

On other hand, elongations during tensile tests show a progressive increase with the Zn addition, while elongations in compression tests did not show any significant changes, see Table 4.2. Pure Mg has a particular behaviour in this respect; elongation in compression test for pure Mg is longer than in tension test. However, this behaviour turns inversely even with an addition of 1 wt. % of Zn.

4. Results

Table 4.2: Mechanical properties of the indirectly extruded pure Mg and Z-series alloys. * Samples showing micro-yielding. $\Delta\sigma$ = TYS – CYS

Alloy	Extrusion speed [m/min]	Tension			Compression			
		TYS [MPa]	UTS [MPa]	A [%]	CYS [MPa]	UCS [MPa]	A [%]	$\Delta\sigma$ [MPa]
Mg	2	37*	212	11	65	337	16	-28
	10	34*	210	12	61	336	14	-27
Z1	1	140	270	21	61	347	14	79
	5	37*	218	15	73	351	13	-36
	10	35*	215	17	57	332	12.5	-22
Z2	1	133	220	25	76	342	13.5	57
	5	56*	218	18	64	333	13	-8
	10	39*	215	19	57	326	13	-18
Z3	1	130	230	26	77	343	15	53
	5	49*	228	22	65	331	14.5	-16
	10	41*	226	22	57	323	13.5	-16
Z4	1	132	242	27	78	377	18	54
	5	83*	236	24	65	357	15	18
	10	41*	234	24	59	338	15	-18

4. Results

4.5.2 ZK-series alloys

Fig. 4.30 (a)-(c) shows the representative stress-strain curves in tension and compression of the ZK-series alloys indirectly extruded at different speeds. In contrast to the previous group of alloys, where perfect parabolic hardening was observed during tension tests, the ZK-series alloys show linear behaviour, indicating that plastic deformation during tensile testing has not taken place homogeneously. On other hand, the flow curves in compression show the same sigmoidal hardening associated with twinning-dominated mechanisms. It is interesting to note that the sharp yield point developed during compression testing becomes smoother with Zn additions. This is especially remarkable in the alloys extruded at higher speeds.

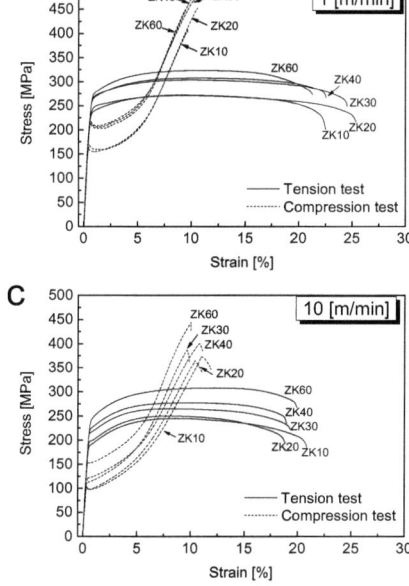

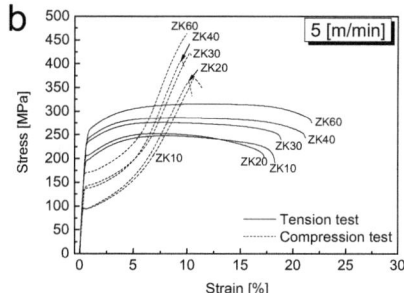

Fig. 4.30. Flow curves in tension and compression for the indirectly extruded ZK-series alloys at speeds of 1 (a), 5 (b) and 10 (c) [m/min].

Fig. 4.31 (b) shows the tensile and compressive yield strength for the indirectly extruded ZK-series alloys. With the addition of Zr, both the tensile and compressive yield strength increase. Also an increase was observed with increasing Zn content. This strengthening effect is strongly reduced when the extrusion speed is increased. This change is connected to the coarser grain sizes developed at higher extrusion speeds, see Figs. 4.16-4.18. However this decrease in the strength is more dramatic for the CYS than for the TYS; thus the yield strength asymmetry is accentuated.

4. Results

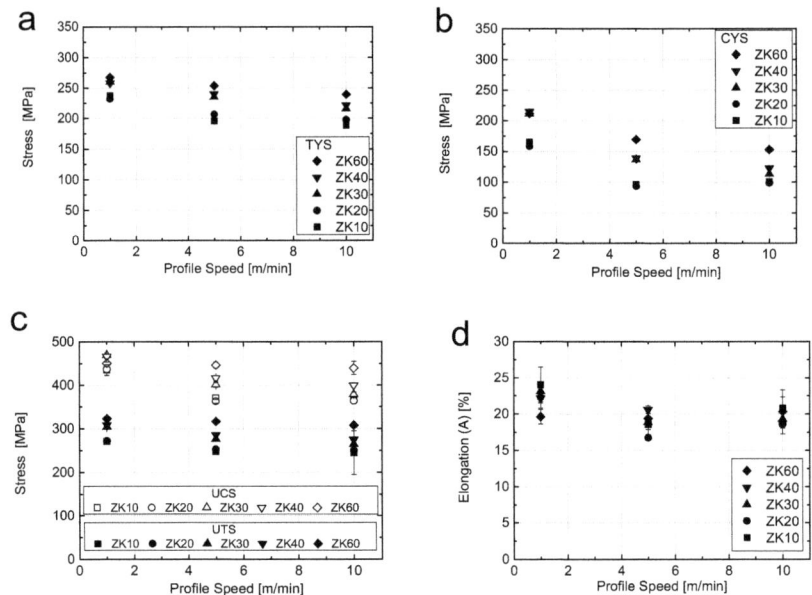

Fig. 4.31. Mechanical properties of the extruded ZK-series alloys: (a) Tensile yield strength TYS, (b) Compressive yield strength CYS, (c) Ultimate tensile strength UTS and Ultimate compressive strength UCS, and (d) Elongation A.

The UCS values are higher than the UTS values, as observed in the binary alloys; see Fig. 4.31 (c). However, the UTS and UCS are increased considerably and additionally they show a progressive increase with Zn additions.

The elongations during tension tests of this series are shown in Fig. 4.31 (d). In contrast to the pure Mg and Z-series alloys, the ZK-series alloys do not show any significant effect of the Zn addition on the elongations. Only alloys extruded at 1 [m/min] showed a slight improvement in this property in comparison with those extruded at higher rates. On the other hand, elongations during compression tests of this series also showed reduced values with a constant value of ~10 %, for the all alloys in all conditions. These and the rest of the numerical values are summarised in Table 4.3.

Table 4.3: Mechanical properties of the indirectly extruded ZK-series alloys.

Alloy	Extrusion speed [m/min]	Tension			Compression			Δσ [MPa]
		TYS [MPa]	UTS [MPa]	A [%]	CYS [MPa]	UCS [MPa]	A [%]	
ZK10	1	237	271	24	165	436	10	72
	5	196	248	19	96	370	11	100
	10	188	245	21	101	366	12	87
ZK20	1	232	272	22	158	437	10	74
	5	207	253	17	94	362	11	113
	10	198	250	19	99	364	11	99
ZK30	1	260	304	23	211	470	10	49
	5	235	276	19	137	401	9	98
	10	216	265	20	113	379	10	103
ZK40	1	258	308	22	215	466	10	43
	5	240	286	21	139	416	10	101
	10	222	276	19	123	340	10.5	99
ZK60	1	267	323	20	212	451	10	55
	5	254	317	19	169	447	9	85
	10	240	308	20	153	440	9	87

4. Results

4.5.3 ZEK-series alloys

Fig. 4.32 (a)-(c) shows the representative stress-strain curves in tension and compression of the ZEK-series alloys extruded at different speeds. For the tensile flow curves, a marked change from a non-hardening behaviour of the alloys extruded at 1 [m/min] to an intermediate between parabolic hardening and linear hardening shown by the alloys extruded at higher rates can be seen. This plastic behaviour is even more inhomogeneous than that observed for the ZK-series alloys.

In contrast to the compression flow curves shown by the other series, sharp yield points are very pronounced and the sigmoidal hardening even more marked.

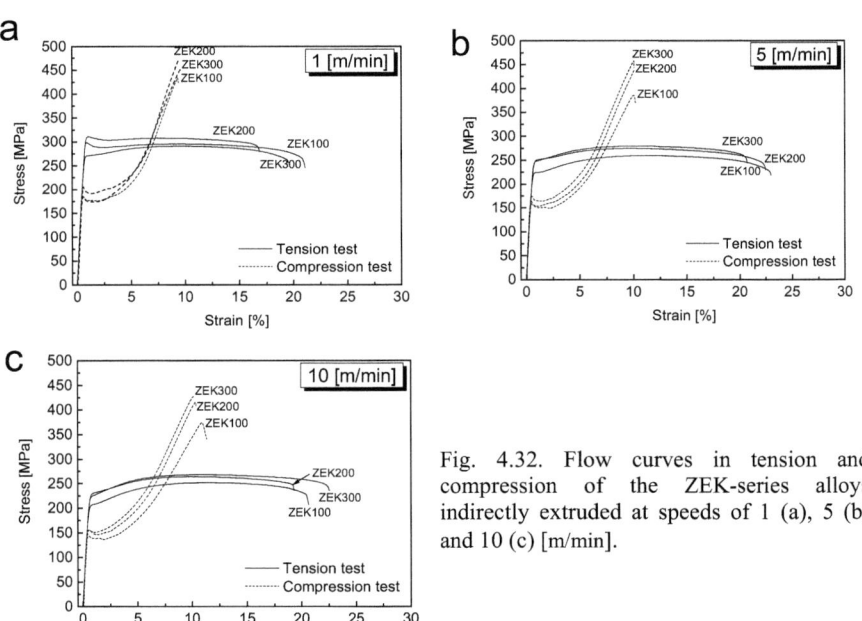

Fig. 4.32. Flow curves in tension and compression of the ZEK-series alloys indirectly extruded at speeds of 1 (a), 5 (b) and 10 (c) [m/min].

With further addition of RE to the ZK-series, there is a further increase in the tensile-compressive yield strength; see Fig. 4.33 (a) and (b). Thus, ZEK series showed the highest values of yield strength of all the alloys studied in the present work. In contrast to their counterparts without RE, the progressive increase in Zn content did not show any proportional increase in the yield strength. However the alloy with 2 wt.% Zn showed the highest TYS and

4. Results

CYS values of all the alloys. The ZEK-series alloys extruded at higher speeds showed a reduction in yield strength. This change is also connected to the coarser grain sizes developed at higher extrusion speeds; see Figs. 4.21 and 4.22. However the yield asymmetry was also somewhat reduced, see Table 4.4.

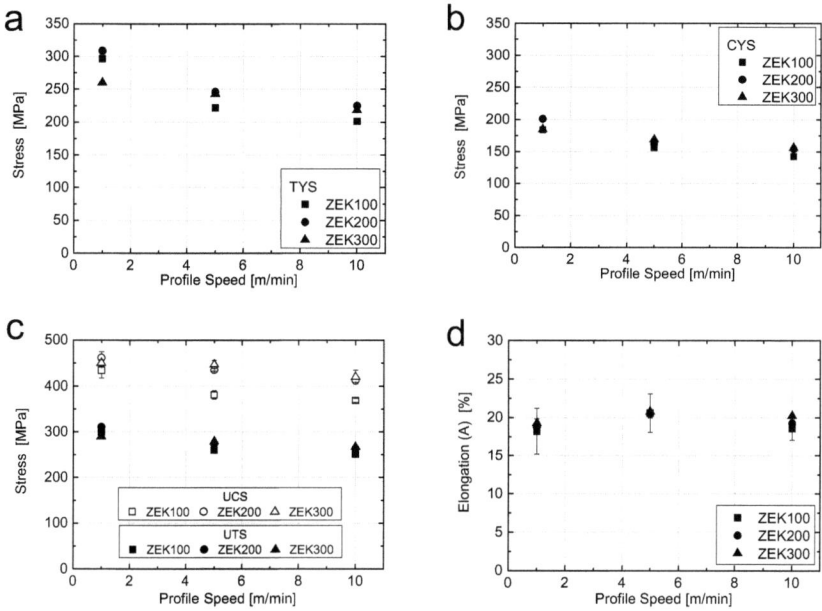

Fig. 4.33. Mechanical properties of extruded ZK-series alloys: (a) Tensile yield strength TYS, (b) Compressive yield strength CYS, (c) Ultimate tensile strength UTS and Ultimate compressive strength UCS, and (d) Elongation A.

Similar to the rest of the alloys, UCS values are also higher than those of UTS; see Fig. 4.33 (c). Compared with the values for similar Zn contents in the ZK-series, the UTS and UCS increase slightly. Zn addition in the ZEK-series shows a slight increase in the UTS, whereas a negligible effect is observed in the UTS.

Elongations also did not show any change with the Zn addition. Moreover elongations achieved in tension tests for all the extrusion conditions showed a *quasi*-constant value of ~ 20 %, see Fig. 4.33 (d). Fracture strains during compression testing also show a constant value of ~ 10 %, see Table 4.4.

4. Results

Table 4.4: Mechanical properties of the indirectly extruded ZEK-series alloys.

Alloy	Extrusion speed [m/min]	Tension			Compression			Δσ [MPa]
		TYS [MPa]	UTS [MPa]	A [%]	CYS [MPa]	UCS [MPa]	A [%]	
ZEK100	1	296	299	18	184	434	9	112
	5	221	260	21	156	381	10	65
	10	201	251	19	142	369	11	59
ZEK200	1	308	311	19	201	462	9	107
	5	246	275	20	162	435	10	84
	10	225	264	19	154	412	10	71
ZEK300	1	260	290	19	185	450	9.5	75
	5	243	279	21	169	447	10	74
	10	218	267	20	156	420	10	62

4. Results

4.6 Annealing treatments

In order to study the progress of recrystallisation, extruded profiles of the ZK and ZEK-series alloys were annealed giving respect to the partly recrystallised condition of these alloys after extrusion. The purpose of these experiments was to produce significant microstructural changes related to static recrystallisation. Thus, the microstructures and textures of these annealed alloys may provide more information about the development of this mechanism. It is important to mention that a first attempt, using an annealing treatment of 12 hours at 200 °C was carried out. However, the resulting microstructures and textures did not show any significant changes. Thus considering the exponential character of the thermally activated recrystallisation mechanisms, an annealing treatment at higher temperature for shorter times was suggested. Thus an annealing treatment of 1 hour at 400 °C was carried out. This treatment produced more significant changes in a shorter time. In the following, the microstructures and textures of the ZK20, ZK40 and ZK60 alloys (representing the side rich in Zn of the ZK series) indirectly extruded at 1 and 10 [m/min] after an annealing treatment of 1 hour at 400 °C are shown.

4.6.1 Indirectly extruded ZK-series alloys after annealing for 1 hour at 400 °C

Figs. 4.34 and 4.35 show the microstructures of the ZK-series alloys extruded at 1 and 10 [m/min] before and after annealing. The microstructures of the alloys extruded at 10 [m/min] did not show any significant changes after annealing, whereas the microstructures of the alloys extruded at 1 [m/min] changed considerably. In the case of the alloy ZK20, most of the large unrecrystallised grains observed in the extruded condition disappeared and additionally grain coarsening can be observed. In contrast, the ZK40 and ZK60 alloys still show large fractions of large grains surrounded by newly recrystallised grains.

The textures of the annealed alloys show some changes which are correlated with those changes observed in the microstructures; see Figs. 4.36 and 4.37. For example, the texture components of the alloys extruded at 1 [m/min] showed drastic changes: i.e., the ZK20 alloy showed complete migration of the strong texture component from the $\langle 10\bar{1}0 \rangle$ pole towards the $\langle 11\bar{2}0 \rangle$ pole (see Fig. 4.36 (a)-(b)), whereas the ZK40 and ZK60 alloys showed a change from a previously strong component focussed on the $\langle 10\bar{1}0 \rangle$ pole towards a double-texture component distributed between the $\langle 10\bar{1}0 \rangle$ and $\langle 11\bar{2}0 \rangle$ poles; see Fig. 4.36 (c)-(f). On other

4. Results

hand, the texture components found in the alloys extruded at 10 [m/min] did not show any significant change with respect to the extruded condition; see Fig. 4.37 (a)-(f).

The changes in intensity with respect to those in the extruded condition have been added as gradients of intensity (ΔI) to Figs. 4.36 and 4.37. In general, all the alloys show a decrease in the intensity ($-\Delta I$); however, for the alloys extruded at 1 [m/min], ΔI is more significant than for those extruded at 10 [m/min]. The highest gradient corresponds to the alloy ZK20, which is correlated with the most advanced recrystallisation process observed in the microstructure. On other hand, the intensity of the $\langle 11\bar{2}1 \rangle$ texture component found in the alloys ZK40 and ZK60 in the extruded condition has also decreased.

If the effect of the addition of Zn is observed separately, it is clear that this inhibits the displacement of the texture component from the $\langle 10\bar{1}0 \rangle$ pole towards the $\langle 11\bar{2}0 \rangle$ pole, similar to the behaviour in the extruded condition.

4. Results

1 [m/min]

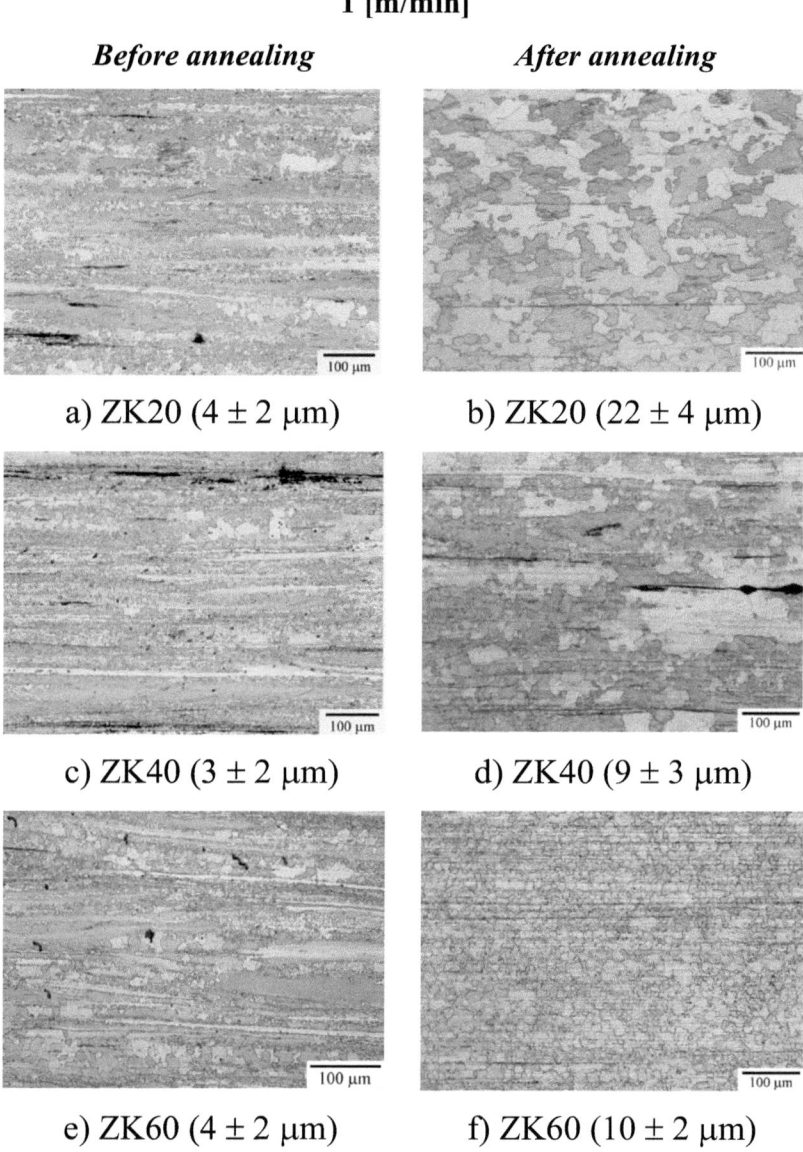

Fig. 4.34. Microstructures of profiles indirectly extruded at 1 [m/min] before and after an annealing treatment at 400 °C for 1 hour: a) and b) ZK20, c) and d) ZK40, e) and f) ZK60.

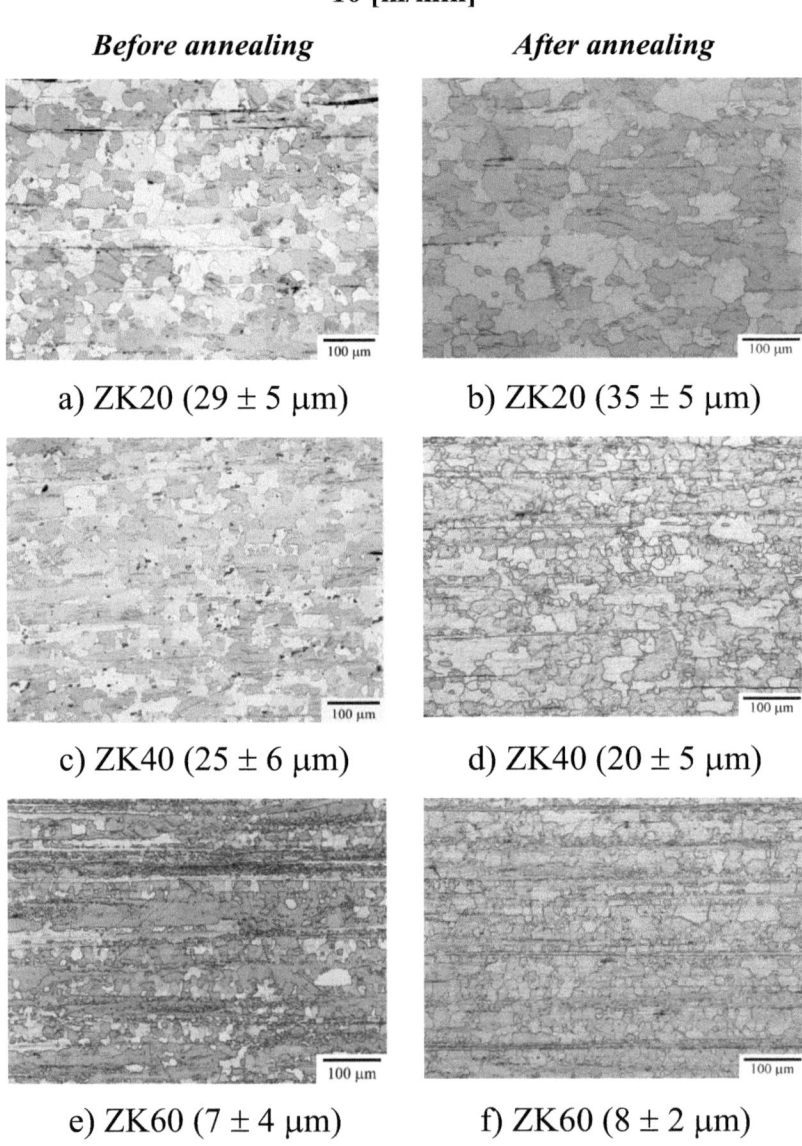

Fig. 4.35. Microstructures of profiles indirectly extruded at 10 [m/min] before and after an annealing treatment at 400 °C for 1 hour: a) and b) ZK20, c) and d) ZK40, e) and f) ZK60.

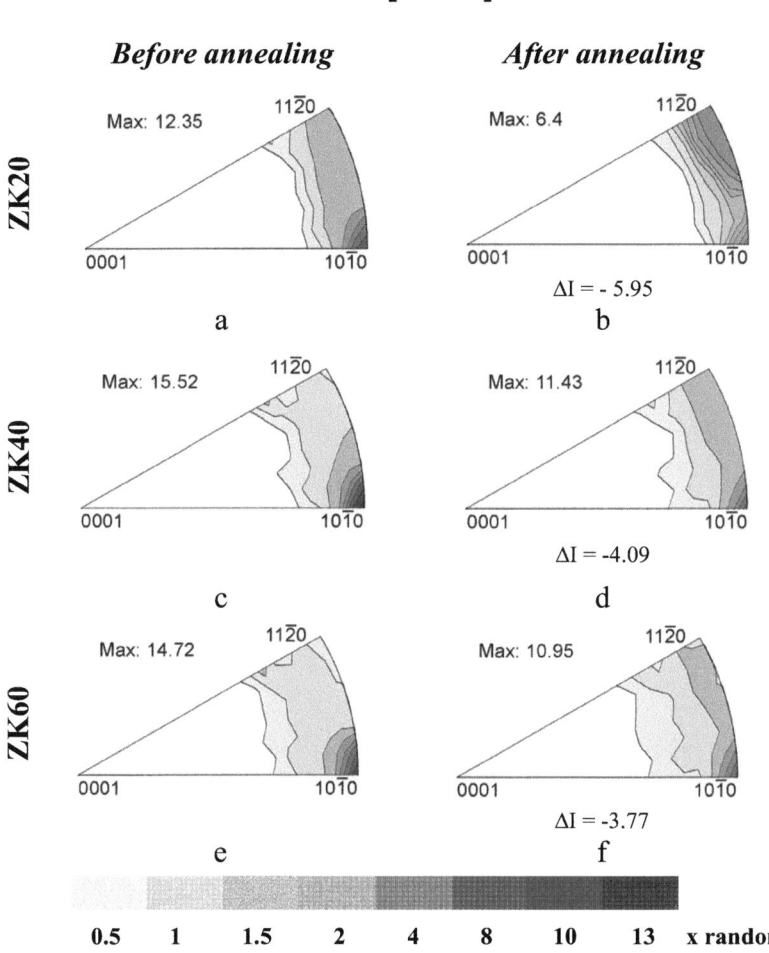

Fig. 4.36. Inverse pole figures of profiles indirectly extruded at 1 [m/min] before and after an annealing treatment at 400 °C for 1 hour: ZK20 (a)-(b), ZK40 (c)-(d) and ZK60 (e)-(f). ΔI (gradient of intensity) is the change in the maximum intensity with respect to that measured in the extruded condition.

4. Results

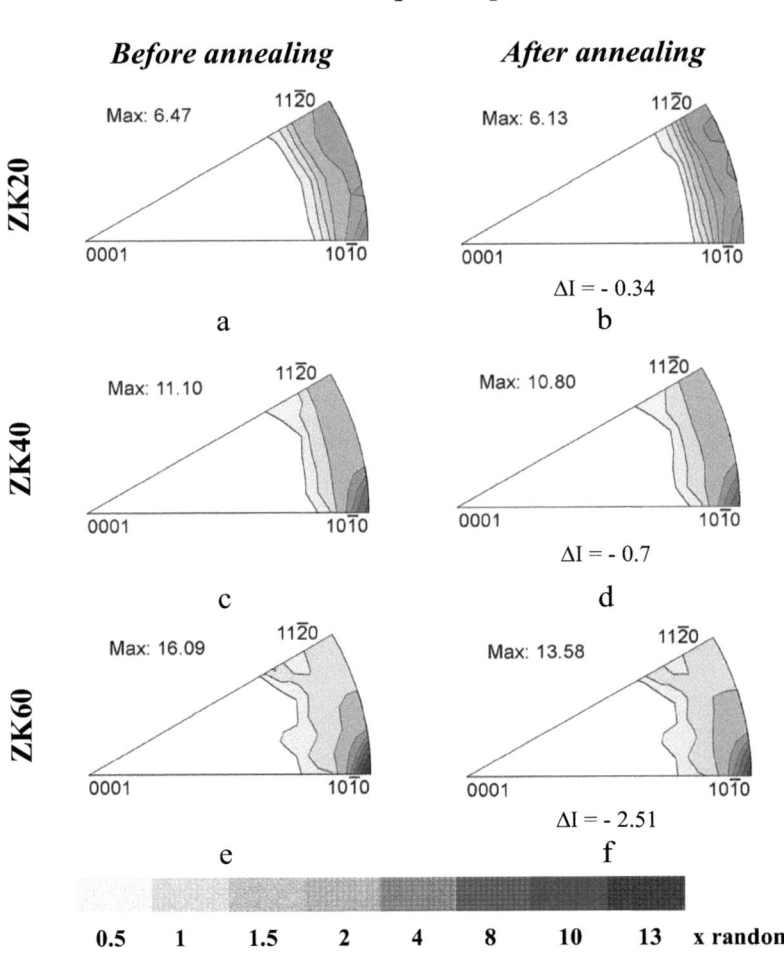

Fig. 4.37. Inverse pole figures of profiles indirectly extruded at 10 [m/min] before and after an annealing treatment at 400 °C for 1 hour: ZK20 (a)-(b), ZK40 (c)-(d) and ZK60 (e)-(f). ΔI (gradient of intensity) is the change in the maximum intensity with respect to that measured in the extruded condition

4. Results

4.6.2 Indirectly extruded ZEK-series alloys after annealing for 1 hour at 400 °C

Figs. 4.38 and 4.39 show the microstructures of the ZEK-series alloys after the annealing treatment. The appearance of new, recrystallised grains is visible in all the microstructures. Compared with the extruded condition; the alloys extruded at 1 [m/min] (see Fig. 4.38) show the most significant changes: most of the former large unrecrystallised grains observed in the extruded condition now consist mainly of new recrystallised grains.

After annealing, the texture intensities decreased considerably for the alloys extruded at 1 [m/min] (see Fig. 4.40), whereas the intensities of the alloys extruded at 10 [m/min] show a negligible change (see Fig. 4.41). The gradients of intensity (ΔI) are similar in magnitude, independent of the alloy composition, i.e. of the Zn content.

After the annealing treatment these alloys tend to develop a $\langle 11\bar{2}1 \rangle$ texture component rather than the double-texture component observed in the ZK-series alloys. This component is better developed for the alloys extruded at 10 [m/min], where apparently the intensities did not exhibit any change. The position of the $\langle 11\bar{2}1 \rangle$ pole is also marked with black arrows in Figs. 4.40 and 4.41.

1 [m/min]

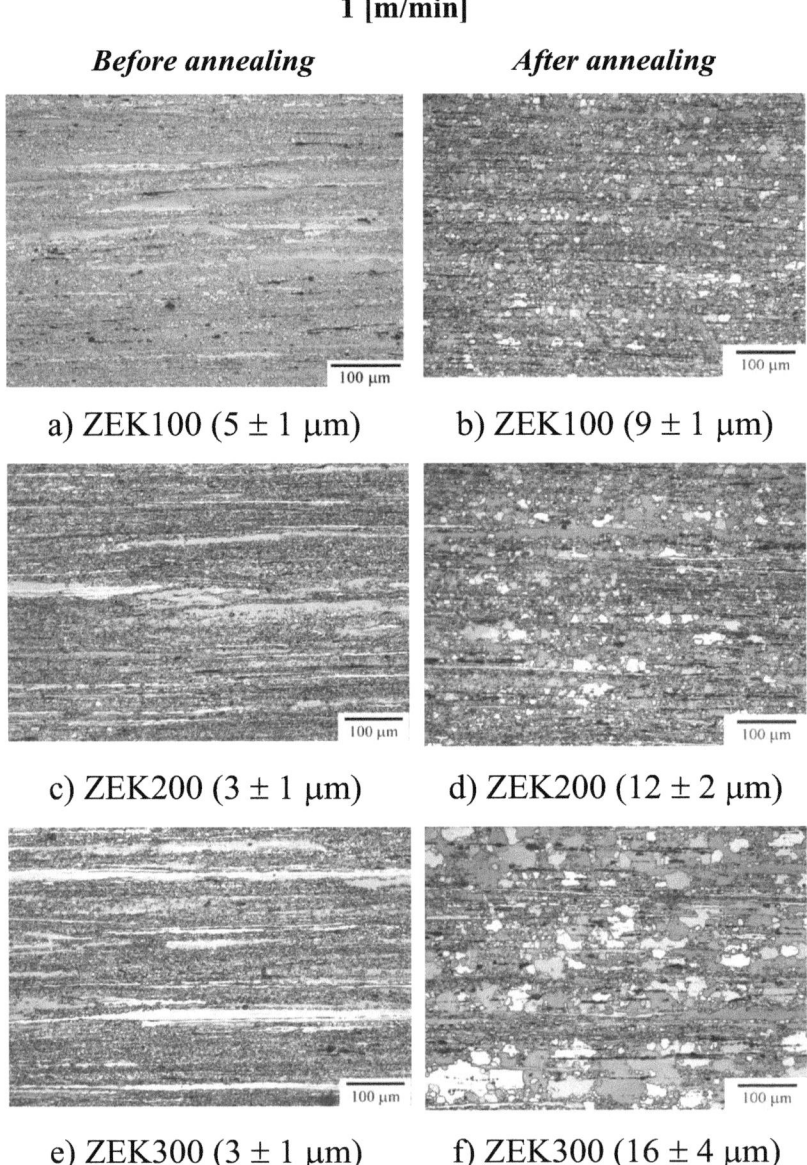

Before annealing   *After annealing*

a) ZEK100 (5 ± 1 µm)   b) ZEK100 (9 ± 1 µm)

c) ZEK200 (3 ± 1 µm)   d) ZEK200 (12 ± 2 µm)

e) ZEK300 (3 ± 1 µm)   f) ZEK300 (16 ± 4 µm)

Fig. 4.38. Microstructures of profiles indirectly extruded at 1 [m/min] before and after an annealing treatment at 400 °C for 1 hour: a) and b) ZEK100, c) and d) ZEK200, e) and f) ZEK300.

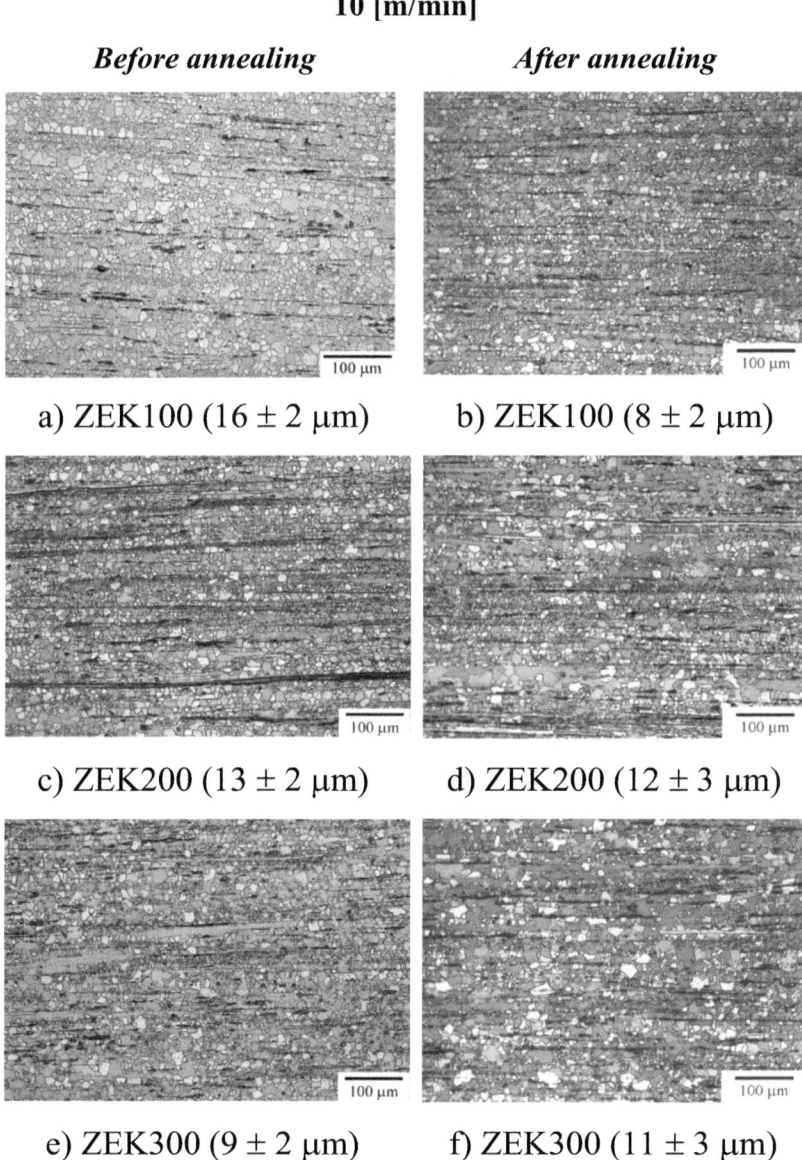

Fig. 4.39. Microstructures of profiles indirectly extruded at 10 [m/min] before and after an annealing treatment at 400 °C for 1 hour: a) and b) ZEK100, c) and d) ZEK200, e) and f) ZEK300.

4. Results

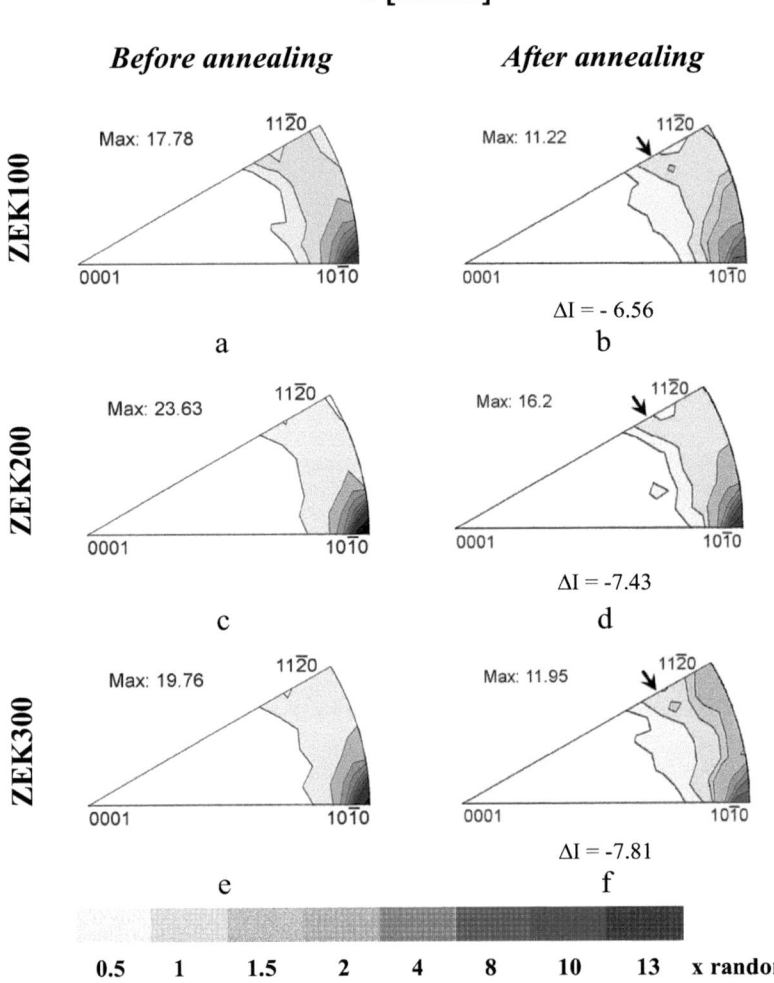

Fig. 4.40. Inverse pole figures of profiles indirectly extruded at 1 [m/min] before and after an annealing treatment at 400 °C for 1 hour: ZEK100 (a)-(b), ZEK200 (c)-(d) and ZEK300 (e)-(f). ΔI (gradient of intensity) is the change in the maximum intensity with respect to that measured in the extruded condition. The position of the $\langle 11\bar{2}1 \rangle$ pole is marked with black arrows.

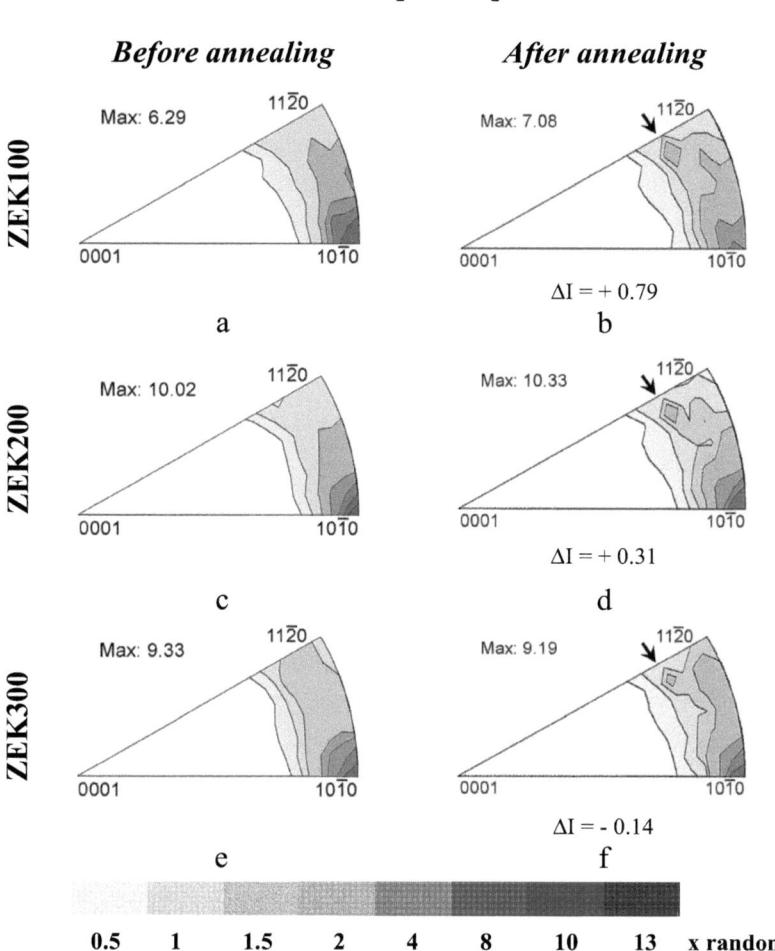

Fig. 4.41. Inverse pole figures of profiles indirectly extruded at 10 [m/min] before and after an annealing treatment at 400 °C for 1 hour: ZEK100 (a)-(b), ZEK200 (c)-(d) and ZEK300 (e)-(f). ΔI (gradient of intensity) is the change in the maximum intensity with respect to that measured in the extruded condition. The position of the $\langle 11\bar{2}1 \rangle$ pole is marked with black arrows.

5. Discussion

5.1 Cast alloys and their influence on the extrusion process

5.1.1 Effect of the alloying element additions on the grain size of the billets

The grain sizes of the homogenised, cast billets summarised in Fig. 4.5 showed a strong influence of the alloy element additions on grain size. Considering that the grain size is determined by both alloy composition and solidification characteristics, then the grain sizes obtained after casting of the alloys in this work are basically influenced by the alloying elements additions, since the casting conditions were the same for all alloys. In particular, it is important to point out that all alloys were poured into preheated steel containers and cooling rates were typically 2-5 °C/s. Under these conditions, a possible grain refinement caused by superheating of the melt is unlikely, since the amount of thermal undercooling required for nucleation (ΔT_n) is dependent on the cooling rate, which can be considered constant for the alloys in this work.

Zn additions in the binary Mg-Zn alloys resulted in grain refinement which is not inconsiderable even with the low Zn contents used, which do not exceed the maximum solid solubility level. In accordance with the eutectic phase diagram of the binary Mg-Zn system and their Zn contents, at the 740°C melting temperature, Zn, with a melting point of 419°C, was completely dissolved in the melt and was kept in solid solution during solidification after pouring. Therefore, the grain refinement effect seems to be correlated with inoculation influenced by the degree of solute segregation, which has been defined by a Growth Restriction Factor Q. This factor is based on a model that has been verified for a broad range of aluminium alloys [108] and has also been applied to some magnesium alloys [4, 46].

The mechanism is based on constitutional undercooling of the melt which is attributed to the presence of solute elements in a diffusion layer ahead of the advancing solid/liquid interface. This restricts grain growth because the diffusion of the solute occurs slowly; thus it limits the rate of crystal growth. In addition, further nucleation occurs in front of the interface (in the diffusion layer) because nucleants in the melt are more likely to survive and be activated in the constitutionally undercooled zone (constitutional undercooling is a major driving force for nucleation).

5. Discussion

The Q factor is a measure of the restriction of the growth velocity and it can be determined for each solute atom from the binary phase diagram using:

$$Q = m_L(k-1)C_0 \tag{5.1}$$

where m_L is the slope of the liquidus line, C_0 is the alloy solute concentration, and k is the equilibrium distribution coefficient = C_S/C_L; where C_S= solute concentration in the solid; C_L= solute concentration in the liquid. This correlation is only valid as long as the solute atom remains dissolved in the melt at dilute concentration levels. The growth restriction factor values for various alloying elements in magnesium alloys have been determined using binary phase diagrams [36, 46] and are shown in Table 5.1. It can be observed that the growth restriction factor for binary Mg-Zn alloys in terms of $m_L(k-1)$, is higher than that of Ce and Al. This value indicates that small additions of Zn should result in effective grain refinement of the alloy, as was verified in the present work.

Table 5.1: Slope of the liquidus line, m_L, equilibrium distribution coefficient, k, and growth restriction factor, $m_L(k-1)$, for some alloying elements in magnesium.

Element	m_L	k	$m_L(k-1)$	System
Zn	-6.04	0.12	5.31	eutectic
Ce	-2.86	0.04	2.74	eutectic
Zr	6.90	6.55	38.29	peritectic
Al	-6.87	0.37	4.32	eutectic

However the importance of Q is also related to the possibility to predict the resulting grain size d from phase diagram parameters through:

$$d = a + \frac{b}{Q} \tag{5.2}$$

where a and b are constants. These constants provide additional information on the potency of nucleant particles, i.e. the slope b, is related to the potency of the nucleant particles where a steeper slope corresponds to a lower potency, and the intercept, a, corresponds to the number of particles that actually nucleate grains at infinite values of Q. For the determination of nucleant particles and their number, a more detailed analysis is required, which is beyond the

5. Discussion

scope of this work. However, additional information can be obtained with the use of the Q relationship and the present results.

Fig. 5.1 shows a plot of grain size versus $1/Q$ for the binary Mg-Zn alloys in this work together with other results for binary Mg-Zn alloys. The values of a and b for all the alloys in Fig. 5.1 are summarised in Table 5.2.

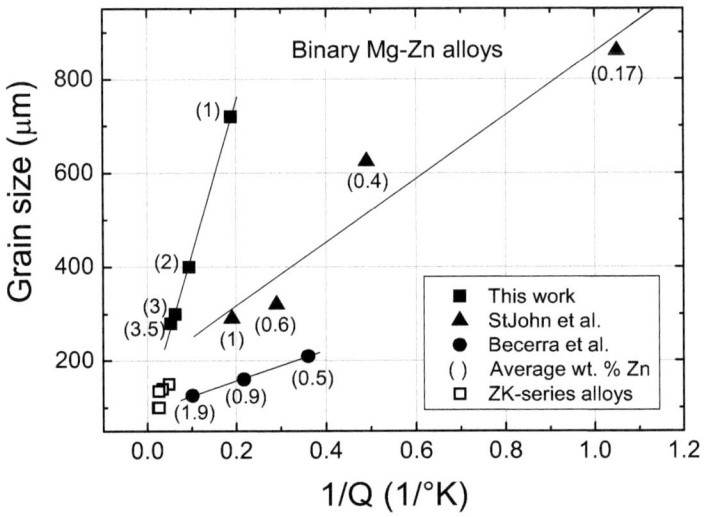

Fig. 5.1. Grain size versus $1/Q$ for Mg-Zn binary alloys.

In Fig. 5.1, it can be noted that grain size is proportional to $1/Q$ for the binary Mg-Zn alloys investigated in this work. This linear relationship is also observed in the results of Becerra [109] and St. John [4]. This confirms the model's potential to predict the grain size as a function of the Zn solute content. The principal differences between the three group of results are in the values of the slope b and the slight deviations in the value of a. As mentioned above, the a value correlates with the number of particles that actually nucleate grains. Since the solute is in all cases Zn, then this number should be the same for all the alloys. This assumption appears to be reasonable as the intercepts, a, are approximately similar. Actually only the a value observed by St. John is considerably different. However this deviation could be attributed to the low Zn contents used in his work, which do not exceed 1 wt. % Zn, thus producing a fluctuating tendency that hinders an adequate calculation of the a value. Impurities could also affect this value.

5. Discussion

On the other hand, the significant variations in the slope *b* indicate that the potency of the nucleant particles is severely affected by differences in the casting conditions, see Table 5.2. Basically, the solidification rates in the referenced works would be faster than that in the present work. Moreover, the volume of the billets cast in this work is considerably larger than that in the other references. A similar effect of cooling rate was observed by Easton et al. [110] for aluminium alloys, where *b* values decreased with increased cooling rate.

Table 5.2: Growth restriction factor relationships and casting conditions.

Mg-Zn alloys	Intercept a (μm)	Gradient b ($\mu m\ K$)	Cast conditions
This work	98.5	3285	Temp. 740°C, poured into steel moulds, Cooling rate = 2-5 °C/s Cylindrical billets: 100 mm Ø x 410 mm
St. John et al.[4]	201	668	Temp. 700°C, poured into steel moulds, Cooling rate = 2-5 °C/s Cylindrical billets: 50 mm Ø x 70 mm
Becerra et al.[109]	94	312	Temp. 680°C, poured into copper moulds, Cooling rate = 38-40 °C/s Discs: 40 mm Ø

The addition of Zr to the binary Mg-Zn alloys used in this work produced further significant grain refinement, regardless of the cast billet volume factor, see Fig. 4.5. Grain refinement by Zr in magnesium is very well established both scientifically and commercially; however the mechanism by which refinement occurs is not very well understood. If the mechanism of growth restriction by solute additions is applied, it is clear that the growth restriction factor, $m_L(k-1)$ for Zr is somewhat larger than that of the other solute elements shown in Table 5.1, and thus it should be more effective. However, the grain refinement mechanism is attributed not only to the soluble Zr content, but also to undissolved Zr particles present in the melt [47, 111], which were observed after casting, see Fig. 4.7. Dissolved and undissolved Zr coexist in the melt due to the high melting point of Zr (1852°C), the low solubility of Zr in molten Mg (only 0.6 wt. %) and the different solidification mechanism via the peritectic reaction in the Mg-Zr binary system. If a Zr content higher than the maximum solubility is added to magnesium (which is usual in these alloys), grain refinement will be further reinforced, as the $m_L(k-1)$ factor indicates. However, the reason for this grain refinement would be attributed not only to the soluble Zr content, but also to the greater number of undissolved Zr particles present in the melt.

5. Discussion

The soluble Zr content in a magnesium alloy is also affected by the alloy composition, e.g. it has been found that the soluble Zr content increases as a function of the Zn content in Mg-Zn alloys [9, 45]. However, Zn and Zr also form stable intermetallic compounds when the Zn content exceeds about 4 wt. % [45]. If such particles form during melting, then they could affect the inoculation of grains during solidification; some Zn-Zr compounds were observed after casting the ZK-series alloys, especially at higher Zn contents, see Fig. 4.7.

In theory, it would be possible to calculate the growth restriction factor in multi-component alloys by conventional summing of the individual Q values for each solute. Considering the points above, the author considers that the grain refinement mechanisms resulting from Zr addition cannot be represented by equations 5.1 and 5.2. Instead, other models that consider most of the above mentioned characteristics of the ternary Mg-Zn-Zr system should be developed. Still, Q values for cast ZK-series were calculated assuming a whole dissolution of Zn and Zr in the melting and added to Fig. 5.1.

It is important to observe in Figs. 4.5 and 5.1 that the addition of Zn in the ZK-series alloys produces a moderate but additional grain refinement, which can be attributed to the Zn solute additions as described. Nevertheless, Zn additions also increase the solubility of Zr in magnesium; thus a higher Zr solute content could also stimulate additional grain refinement.

The addition of Ce-MM to the ternary Mg-Zn-Zr alloys contributed to additional grain refinement. However, the principal grain refinement effect can be attributed to the Zr addition. If the grain sizes of the ZK and the ZEK-series alloys are compared (see Fig. 4.5), it is clear that the addition of Ce-MM leads to an additional grain size reduction of approximately 30 %. However, it is unclear if this is due to a contribution by the solute content of Ce und La, (the principal elements present in Ce-MM), or by the presence of the numerous, complex intermetallic compounds formed during solidification and observed after casting, see Fig. 4.8.

5. Discussion

5.1.2 Effects of alloy composition and microstructure on extrusion processing

One of the main aims of this work is to establish a correlation between the microstructure and composition of the initial feedstock material and the thermo-mechanical response, in terms of the development of the extrusion force and temperature measured during processing. This kind of correlation is important because it could allow prediction of the extrusion response according to the chemical composition or initial grain size (the main variables of the feedstock materials). The temperature and stress developed during extrusion determine the characteristics of the resulting profiles; however, this topic will be discussed separately. To understand the influence of the feedstock material on the process, it is important to try to separate the possible contributions of these parameters on the extrusion process. Fig. 4.10 shows that the profile temperatures and steady state forces vary with respect to the different alloys. Actually, the temperatures and extrusion forces increase progressively with the addition of alloying elements, since the lowest values correspond to those for pure magnesium and the binary Mg-Zn alloys and the highest values to those measured for the Mg-Zn-Zr-Ce-MM system or the ZEK-series alloys.

It is not yet clear if this is solely a result of the chemical composition, since as mentioned above, the cast alloys used in this work showed significant differences in grain size that could also influence the temperatures and forces developed during extrusion.

In order to separate and observe the possible effects of the initial grain size on the measured extrusion parameters, the steady state forces and temperatures have been plotted versus grain size in Fig. 5.2. Grain size is represented by $1/d$ to in order to enable trends to be more easily detected. The different series of alloys are identified by different symbols, which are used to represent a group of alloys extruded at the same speed. The force is a parameter registered during the process (see Fig. 4.9) and its value corresponds specifically to the steady state extrusion force. Fig. 5.2 (a) shows that there is no clear trend between initial grain size and the force, although in general it is possible to observe that for grain-refined alloys higher extrusion forces were necessary. Peak forces at the beginning of extrusion are processing values that should show a better correlation with the initial grain size. However, the indirect extrusion process used in this work was designed to avoid such peaks and achieve uniform deformation immediately. The steady state force represents this uniform deformation stage, which involves a dynamic recrystallisation process; hence this value should be better correlated with the final grain size rather than that of the feedstock material. In fact, this is

5. Discussion

reflected in the fluctuation of the data in Fig. 5.2 (a), which seems to be better correlated with the alloy and its response during deformation rather than with the initial grain size.

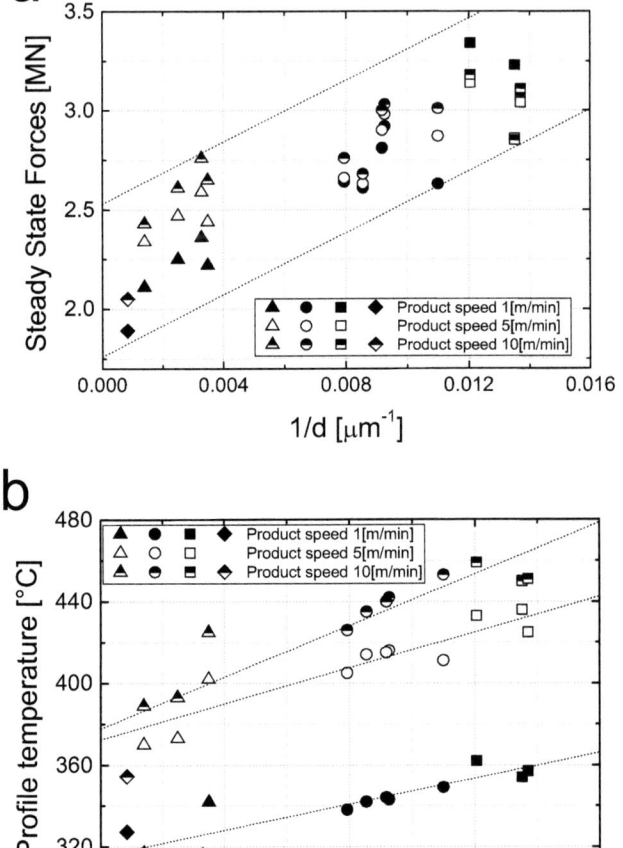

Fig. 5.2. a) Steady state force *vs* Inverse average grain size (1/d), and b) Profile temperature *vs* Inverse average grain size (1/d), Pure Mg* are grouped in diamonds, Z-series alloys in triangles, ZK–series in circles and ZEK –series in squares.* 2 and 10 [m/min].

The profile temperature is a parameter which may be more dependent on the feedstock characteristics, since it represents the accumulated adiabatic heating caused by deformation from the beginning of the process. Fig. 5.2 (b) shows not only a clear trend between grain size

5. Discussion

and temperature, but also a clear dependence on the extrusion speed. On the other hand, this plot allows us to separate the effects of the alloying element additions. A good example is the case of the ZK-series alloys. These alloys showed similar grain sizes in the as-cast and homogenised condition. Therefore, if grain size were to be the only factor affecting profile temperature then this should be constant for all the alloys of this series. However, the profile temperature increases slightly but progressively with the Zn addition. This tendency is also shown by the ZEK-series alloys, where the same analogy of similar grain size can be applied. Thus a linear trend is clearly observed in the 1/d plots, with the slopes appearing to be specific for each extrusion rate.

There exists only a slight fluctuation in the values for pure Mg and the Z-series alloys; however these values still show the same trend. It is important to notice that the scatter in the data at higher extrusion speeds can only be attributed to an alloy dependent deformation response. This response is, as with the extrusion force, better correlated with the microstructure of the resulting profile than with the initial conditions.

In spite of some scatter in the data seen in Fig. 5.2 (b), it is possible to establish a very well defined interdependence of the grain size and profile temperature (linear if is plotted against 1/d) and the scatter in the data observed in this plot seems to be influenced by a different deformation response attributed exclusively to alloy composition.

On other hand, Fig. 5.2 (b) also shows that profile speed and profile temperature are process parameters which maintain a fixed relationship which is represented by a constant slope at each extrusion speed. This plot also indicates that profile temperatures rose significantly when the extrusion speed is increased from 1 to 5 [m/min]; however, only a slight increase in profile temperature is observed if the extrusion speed is increased from 5 to 10 [m/min]. Figs. 5.2 (a) and (b) already suggest a possible correlation between locally measured extrusion parameters (temperature and force) and the fixed extrusion parameter (speed).

In order to analyse a possible relationship between the above mentioned process parameters, profile temperature is plotted as a function of extrusion force in Fig. 5.3 for all the values measured in this work. Although the data show some scatter, it is clear that profile temperature and steady state force maintain a linear relationship, furthermore each group of alloys extruded at a fixed extrusion speed shows a linear tendency with a constant slope and

5. Discussion

apparently the three different sets of data show a constant value in the intersection with the y axis (profile temperature) at approximately 300 °C, which is the initial temperature. It is noteworthy that Fig. 5.3 indicates that for the indirect extrusion press used, it is possible to predict the resulting profile temperature and extrusion force when the extrusion speed is kept constant. In agreement with Fig. 5.2 (b) the grain size seems to influence the profile temperature to a higher extent than the alloy composition, thus for a specific alloy with a known initial grain size, it is possible to estimate the resulting profile temperature. As profile temperature has a linear relationship with the resulting steady state force, this value can also be estimated if a constant extrusion speed is assumed. The prediction of these values is important because they determine the resulting microstructure of the extruded profiles in terms of grain size and texture as will be discussed in the next section.

Based on the behaviour shown in Figs. 5.2 (a) and (b), it can be concluded that profile temperature is a value strongly influenced by grain size, and in contrast to this, the steady state force is a value strongly correlated with alloy composition, since it is the response of the alloy to deformation at high temperatures, i.e. it is related to dynamic recrystallisation. These differences are the principal reason for the scatter in the data, which is very accentuated for the data shown in Fig. 5.2 (a).

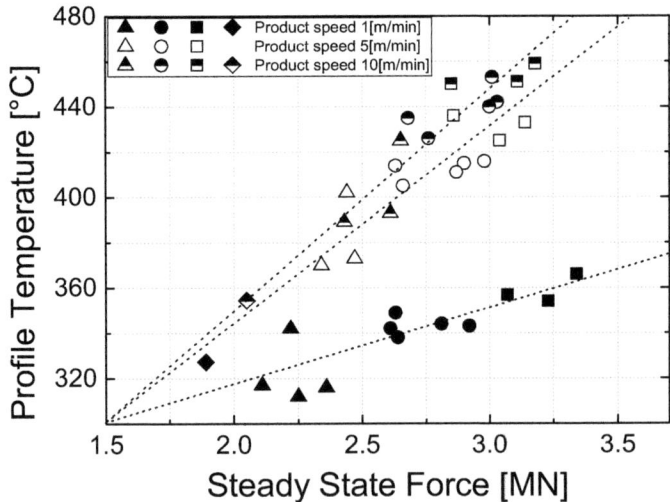

Fig. 5.3. Profile temperature *vs* steady state force. Pure Mg* are grouped in diamonds, Z-series alloys in triangles, ZK –series in circles and ZEK – series in squares.* 2 and 10 [m/min].

5. Discussion

5.2 Correlation between extrusion process variables and the microstructure of the extruded profiles

It has been shown that locally measurable extrusion process parameters are influenced by the characteristics of the cast and homogenised billets. However, these parameters are also closely related to the deformation response of the alloy and the resulting microstructure. In this section an analysis of the thermal and mechanical response of the different alloys to deformation during indirect extrusion is described. Subsequently, the resulting grain sizes and textures of the extruded profiles will be directly correlated with this analysis.

5.2.1 Deformation response of the studied alloys during indirect extrusion

Numerous previous studies investigated the deformation behaviour of magnesium alloys at elevated temperatures are carried out experimentally through compressive or torsion mechanical tests and then these results are used under adequate assumptions to predict their deformation behaviour during thermo-mechanical process. However, strain rates and stresses applied during experiments are usually lower than those actually applied e.g. during extrusion process. A direct measure of variables such as strain, strain rate and billet temperature are extremely difficult to measure on a commercial scale press and nowadays only few data of these characteristics are available. The purpose of this introduction is to understand the importance of the locally measured values reported in this work. For an analysis of these values it is necessary to define exactly what these locally values represent and which information is able to be determined from them.

Profile temperature has been measured directly in the bearing face of the die. In this point, the material has just crossed the highest critical deformation zone; therefore, the temperature measured in this point represents the highest reached heating temperature as consequence of deformation. A higher temperature than that achieved in this point is definitely discarded, thus the profile temperature represents a reliable value of the true maximum temperature reached by the material. On other hand, steady state force represents a uniform deformation region reached at a specific temperature (above 300°C), where the material can be deformed at constant stress and strain rate due to an equilibrium between work softening and work hardening. This stress-strain behaviour is often considered to be a manifestation of dynamic recrystallisation (DRX). Consequently, values measured during extrusion can be used to correlate them with DRX process developed for each alloy during processing.

5. Discussion

For an adequate correlation of these values with the microstructural evolution, it is necessary to deal with three variables: deformation temperature (profile Temperature), strain rate ($\dot{\varepsilon}_t$) and strain; i.e. mean flow stress (σ_f) determined from the extrusion force in stationary state.

From the logged data, the σ_f [MPa] was calculated using [97]:

$$\sigma_F = \frac{F_{SS}}{C \cdot \ln \varphi \cdot A_0} \qquad (5.3)$$

Where F_{SS} is the steady state force [N], $\ln \varphi$ is the natural logarithm of extrusion ratio, A_0 is the cross sectional area of the stem or billet container [mm^2] and C is a factor that is die specific and includes a shape efficiency factor. If a round bar is being extruded, it can be assumed a C factor of 1.66 [97].

Fig. 5.4 shows plots of the calculated mean flow stress in function of the profile temperature and strain rates for the different alloys studied in this work. Each point represents a fixed strain rate, however, temperature and flow stress vary for the different alloys. Temperature is a variable directly dependent on the strain rate, whereas flow stress is a variable dependent on the temperature; however both of them are alloy dependent too. This dependence of several variables turns our analysis complex; nevertheless, the most significant changes observed in the Fig. 5.4 can be already attributed exclusively to alloy composition. In general, the addition of alloying elements and an increasing strain rate tend to increase temperature and flow stress. This is clearly observed for the binary Mg-Zn alloys and the ternary ZK-series alloys, see Fig. 5.4 (a) and (b). Nevertheless, ZEK-series alloys show an opposite behaviour, the addition of Zn and an increasing strain rate tend to decrease flow stress and temperature, see Fig. 5.4 (c).

To simplify this analysis, it is possible to apply a useful relation that describes the combined temperature and strain rate dependence of flow stress, called the Zener-Hollomon parameter (Z). It may also refer to as a *temperature-modified* strain rate [81], which is defined as:

$$Z = \dot{\varepsilon}_t \exp\left(\frac{Q_A}{RT}\right) \qquad (5.4)$$

5. Discussion

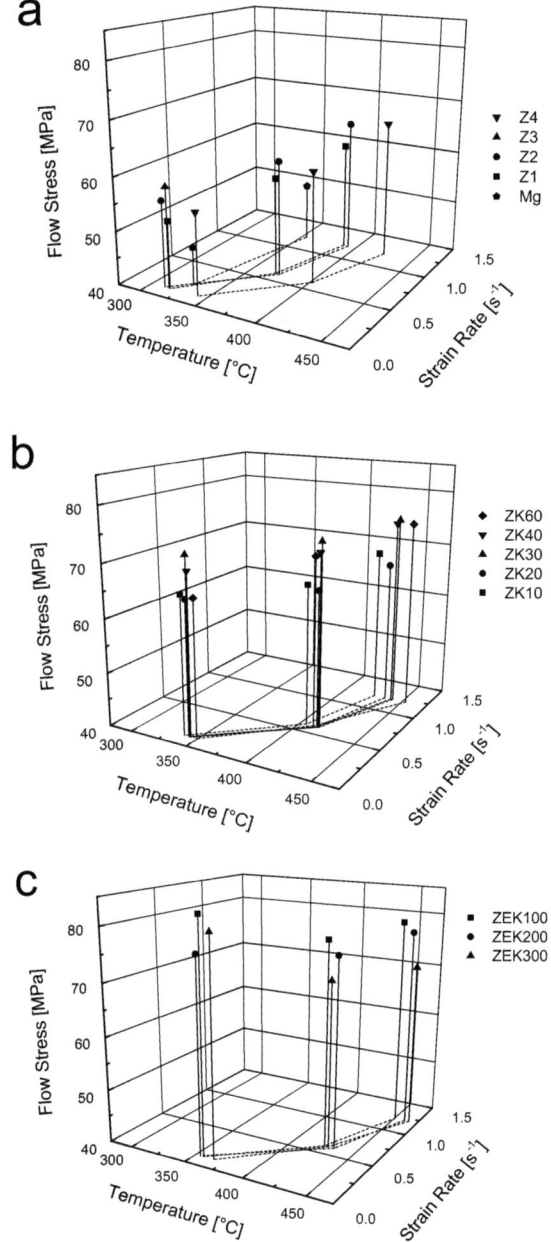

Fig. 5.4. Deformation response of the alloys studied in this work during indirect extrusion. Strain rate was kept constant, whereas temperature and flow stress were independent measured variables.

5. Discussion

where Q_A is the apparent activation energy for deformation, which will be assumed to be similar to that for self-diffusion in magnesium (135000 [J/mol]), R is the universal gas constant (8.314 [J/molK]) and T is the absolute profile temperature [° K].

The Z parameter has been shown to be adequate when describing material behaviour during extrusion processing [112, 113]. In order to find a correlation between Z and the flow stress, both have been plotted for the different groups of alloys in Fig 5.5. In these plots it can be observed that Z varies as a function of the flow stress, a high Z value involves lower stresses (and lower temperatures) and a low Z the opposite. The temperature increased for all extrusion trials with increasing strain rate, thus, the effect of an increasing temperature and strain rate is easily observed from the right side toward the left side in these plots.

Considering an identical initial extrusion temperature and constant strain rates for all the extrusion trials, it is clear that each point in the plot indicates the necessary combination of temperature and strain rate to reach the steady state, where flow stress reaches equilibrium between work hardening and DRX softening. As can be appreciated in Fig. 5.5 alloying elements influence strongly these conditions and each group of alloys respond in different way when the strain rate and temperature increase. If this point is reached at a lower flow stress, it indicates that activation of DRX is easier than in those alloys where higher flow stresses were necessary to reach this equilibrium.

The highest Z values of this work were obtained for pure magnesium and the binary Mg-Zn alloys. With increasing strain rate the flow stress of the Z-series alloys tend to increase. However, at the higher strain rate the flow stress of pure magnesium is decreased.

With the addition of Zr to the binary alloys, the flow stresses and temperatures increased, and thus Z values were lower than those for the binary alloys. Zn additions did not show any significant effect in these alloys. However, a significant reduction of the Z values is observed with increase of the strain rate and temperature.

For the ZEK-series alloys, higher flow stresses were necessary to reach the steady state, thus a larger increase in temperature resulted, giving the lowest values of Z obtained in this work.

5. Discussion

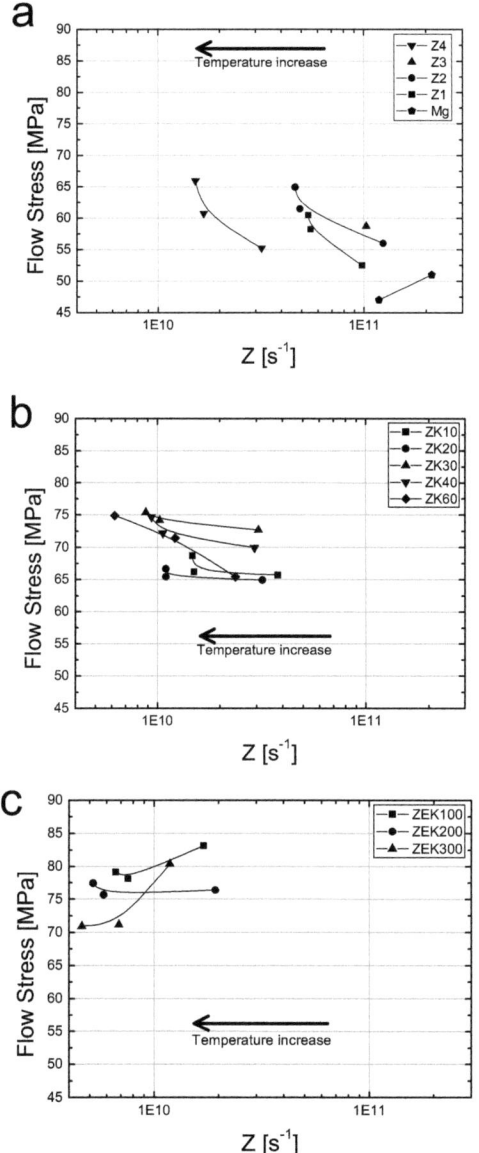

Fig. 5.5. Flow stress versus Z for a) Mg and Z-series, b) ZK-series and c) ZEK-series.

It is noteworthy that at higher strain rates, the flow stress seems to be reduced, which is the opposite effect to that observed in the Z and ZK-series alloys, but similar to that of pure

5. Discussion

magnesium. Similar to the other alloys the Z values are significantly reduced at increased strain rates. The fall in the flow stress and the Z values at higher strain rates can only be correlated with a more effective activation of DRX at the temperatures achieved by these alloys during deformation (430-450°C).

In summary, Z parameter shows the sensibility of the activation of DRX process with the chemical composition; thus Z values have to be correlated also with the microstructural evolution. Z values showed in this section were based on directly measured parameters during extrusion process; therefore, they are reliable data to be considered e.g. in an eventual simulation of the extrusion process of these alloys. Since Q_A values are also alloy dependent, they still need to be determined by each alloy group e.g. by tension or torsion tests.

5.2.2 Relationship between process variables and the resulting microstructure

The importance of the previous analysis of the three principal variables involved in the extrusion process is the direct connection with the resulting microstructure of the extruded profiles. The size of the DRX grains depends strongly on the deformation temperature, therefore, it can be also correlated with the Z parameter and numerous studies have established the dependence of grain size on Z [5, 114, 115]. Fig. 5.6 shows the average grain size of the extruded profiles as a function of the Z parameter. In general it is observed that high Z values are associated with finer grain sizes and low Z values with the opposite. This relationship between grain size and Z can be used to predict the DRX grain size according to the process parameters and alloy composition. In general, higher Z values (low temperatures and strain rates) result in finer grain sizes.

Pure Mg and Z-series alloys

The principal characteristics of the extruded microstructures of the pure Mg and Z-series alloys are their homogeneity and coarse grains. In particular, grain growth in the binary Mg-Zn alloys was significantly enhanced at higher strain rates and temperatures; see Fig. 5.6 (a). These differences can also be correlated with the deformation response observed in Fig. 5.5 in terms of Z, and thus with different kinetics of DRX with respect to the other groups of alloys.

Pure magnesium and the binary Mg-Zn alloys show high Z values; therefore DRX activation for these alloys is easier than for the others, considering similar extrusion parameters.

5. Discussion

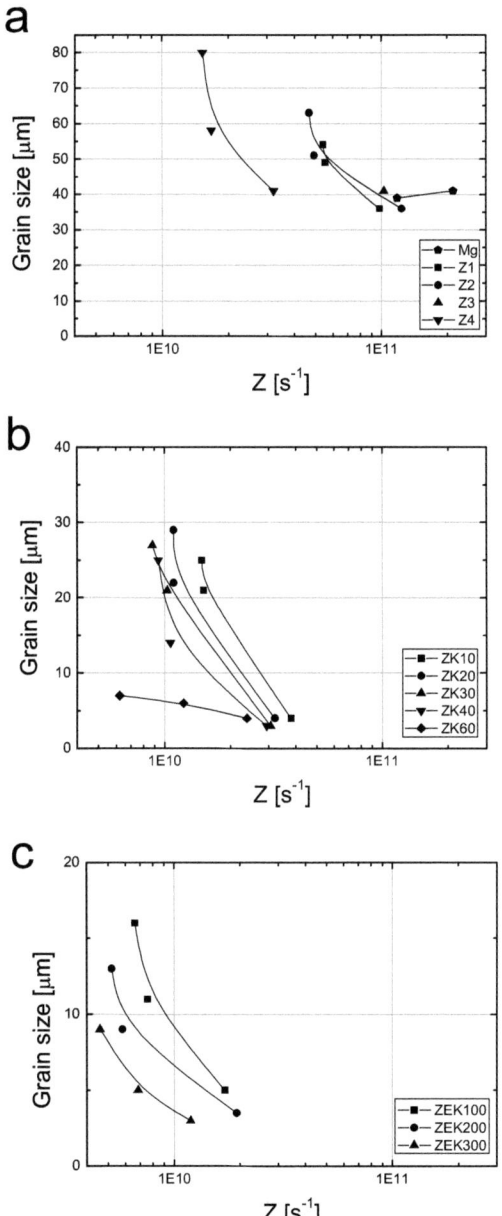

Fig. 5.6. Average recrystallised grain size as function of Zener-Hollomon parameters for: a) pure Mg and Z-series, b) ZK-series and c) ZEK-series.

5. Discussion

This is clearly reflected by the resulting homogeneously recrystallised microstructures of the whole Z-series alloys and pure Mg. Nevertheless, the binary alloys were subjected to higher temperatures (between 370 and 410°C) than pure Mg (maximum 350°C) as a consequence of the higher strain rates.

The fact that pure Mg extruded under similar conditions did not show grain growth indicates that this is an effect attributed exclusively to the Zn addition, as a solute element. This is important because at temperatures above 300 °C, the binary Z-series alloys are located in the solid solution region according to the phase diagram (see Fig. 2.4). It is important to notice that in the solid solution region *solidus* and *liquidus* decrease significantly with the Zn addition. Therefore the melting temperature and the homologous temperature decrease proportionally.

With a decrease in the homologous temperature, thermally activated mechanisms such as diffusion, are more easily activated. It is generally accepted that the grain growth mechanism is controlled by the diffusion of atoms across the grain boundaries. According to Humphreys [81] boundary mobility is controlled by solute concentration. At high solute concentrations, the mobility is low and decreases with increasing solute concentration; however the boundary velocity is controlled by the rate of diffusion of the solute atoms. Thus the combined effects of the temperature and solute on the mobility have an opposite effect producing a sharp increase in mobility at a higher temperature. Experimental data on the diffusion of Zn solute in Mg (measured in the range 327 to 426 °C) showed an exponential increase with the temperature [116]. Additionally, preferential segregation of solute atoms to grain boundaries increases the mobility of the boundary. This segregation of Zn to grain boundaries can be confirmed by the presence of some MgZn intermetallic precipitates lying along the grain boundaries after the extrusion process, see Fig. 4.15. Concerning the precipitates observed in this series, their amount was definitely small compared with the other series of alloys. Therefore they do not present enough obstacles to hinder grain growth. This is confirmed by the homogenous grain growth observed in the microstructure, which indicates that this mechanism is carried out continuously.

ZK- and ZEK-series alloys

Although the grain sizes of the extruded ZK and ZEK-series alloys are significantly finer than those of pure Mg and the Z-series alloys, these alloys were characterised by inhomogeneous

5. Discussion

microstructures in which large unrecrystallised grains coexist with new DRX grains. Similar to the binary alloys, the microstructures of the ZK and ZEK-series alloys also show a strong correlation with the deformation response in terms of the Z values presented in Fig. 5.5. The lower Z values obtained for these alloys indicate that higher stresses and temperatures were necessary to activate DRX. This difficulty in activating DRX is reflected in the inhomogeneous microstructures of the extruded profiles, especially at low temperatures and strain rates. With the increased temperature at higher strain rates, the DRX process was encouraged and the resulting microstructures become more homogenous. Higher temperatures also promoted grain growth in these (also Zn containing) alloys, although this effect is considerably inhibited by the subsequent addition of Zr and Ce-MM

This grain growth inhibition can be attributed to the presence of larger amounts of intermetallic precipitates in these alloys; see Fig. 4.20 and 4.24. Most of these precipitates were already present in the as-cast condition, see Figs. 4.7 and 4.8. Therefore, they could also be influential in retarding DRX, however this will be discussed in the next section together with the resulting textures.

The mechanism by which particles can inhibit grain growth is known as Zener pinning and it is widely used to control the grain sizes of commercial alloys of aluminium and steels [81]. A considerable amount of well distributed particles in a matrix, should act to prevent the motion of the grain boundaries by exerting a pinning pressure, which counteracts the grain growth observed in the binary alloys.

With the addition of Zr to the binary Z-series alloys an increased amount of intermetallic precipitates was observed in the as-cast condition, see Fig. 4.7. The amount of precipitates was increased significantly with the Zn addition. Moreover, this series was the only one where both intergranular and intragranular precipitates were formed as a result of the extrusion process (principally of the intermetallic phase Zn_2Zr), thus satisfying the requirements for Zener pinning. SEM micrographs of the extruded ZK-alloys show that grain growth was considerably inhibited in the alloys with a higher fraction and better distribution of precipitates see Fig 4.20.

Zr and Ce-MM additions to the binary Mg-Zn alloys led to the greatest amount of precipitates observed in the matrix of all the alloys. These were already present in the as-cast condition

5. Discussion

and were broken and became uniformly distributed along large bands during extrusion. Thus, grain growth inhibition during extrusion was more effective. In contrast with the ZK-series alloys, Zn additions did not produce a progressive increase in the amount of precipitates, neither before nor after extrusion. Moreover, the ZEK-series alloys did not develop new precipitates after extrusion. Thus, the composition of the precipitates after extrusion is the same as that observed in the cast and homogenised condition.

This uniform distribution of intermetallic particles in the ZEK-series alloys was fundamental in inhibiting grain growth in the extruded profiles, since these alloys experienced significantly higher temperatures during extrusion (up to 450°C). Moreover, in agreement with the Z values, high temperatures are also required to activate DRX during extrusion of these alloys. Grain growth inhibition in Ce-MM containing wrought magnesium alloys has also been reported by other authors [56, 57].

5.3 Recrystallisation as result of the extrusion processing (DRX)

The DRX mechanism in magnesium and magnesium alloys in the range of temperatures measured in this work (between 300 and 450 °C) involves bulging of the grain boundaries, where subgrain growth takes place and new grains are formed at the bulges of the original grain boundaries. However, the mechanism and/or kinetics of DRX can be influenced by alloy composition as well by the presence of intermetallic particles. In the present work, significant changes in the evolution of the DRX process as a function of the alloying element contents and the process variables were observed. In the following, the microstructures and textures will be correlated in order to discuss the role of alloy composition and deformation response during extrusion (Z values) on DRX. This discussion is based on the macro-texture measurements shown in the results section; some EBSD measurements on selected samples have been included in order to clarify micro-texture details.

Pure magnesium and Z-series alloys

The textures of indirectly extruded pure Mg were characterised by distributions of intensity between the $\langle 10\bar{1}0 \rangle$ and $\langle 11\bar{2}0 \rangle$ prismatic poles with a maximum intensity at around 15 ° from the $\langle 11\bar{2}0 \rangle$ prismatic pole. This tilting was only observed in pure Mg, but it is not exceptional. Actually, it is a rule that pure Mg after extrusion or rolling develops a splitting of basal texture [25, 117]. As consequence of this splitting, prismatic pole are also tilted. This characteristic texture of pure Mg is generally attributed to deformation dominated by tensile

5. Discussion

or compression twins together with basal slip [117]. The microstructure of pure Mg was very homogeneous in all conditions, see Fig. 4.11, which indicates that the recrystallisation process has finished. The coarse grains (~40 μm) observed in these samples confirm this point, since grain growth can only take place after the recrystallisation process has finished [76]. The $\langle 11\bar{2}0 \rangle$ texture component represents the conventional extrusion texture of fully recrystallised, pure magnesium and its alloys [90, 105]. Its origin will be better understood following the analysis of texture development of the other alloys. Nonetheless, this texture can be used to benchmark the effect of the other alloying elements.

The textures of the Z-series alloys were also characterised by intensities distributed between the $\langle 10\bar{1}0 \rangle$ and $\langle 11\bar{2}0 \rangle$ prismatic poles, see Fig. 4.25. Basically they are not so different from those of pure Mg, because they also represent a fully recrystallised state. Although, the $\langle 11\bar{2}0 \rangle$ texture component is better outlined, only slight tilting is observed in those alloys extruded at low speeds. The intensity distribution between the $\langle 10\bar{1}0 \rangle$ and $\langle 11\bar{2}0 \rangle$ poles from inverse pole figures can be used to summarise the variations of texture with extrusion speed, see Fig. 5.7. As a consequence of the increased temperature at higher extrusion speeds and an encouraged DRX process, the intensities on the $\langle 11\bar{2}0 \rangle$ pole were clearly increased. This plot is useful to compare changes between the different alloys, square and circle symbols indicate alloys with low Zn contents and triangle symbols the opposite. The triangle on the right indicates the selected angle in the inverse pole figures shown in the results section.

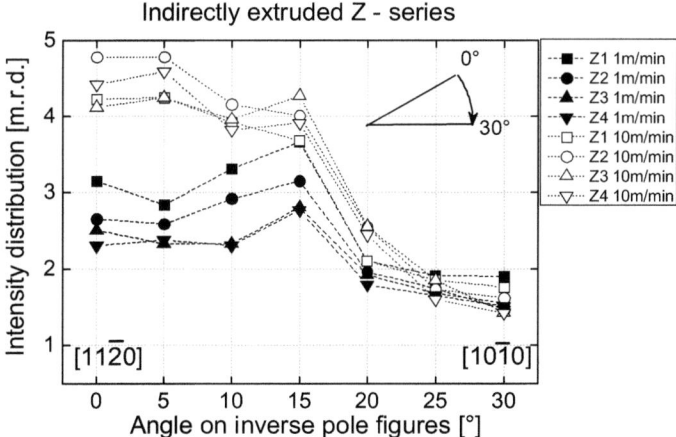

Fig. 5.7. Intensity distribution on the inverse pole figures in the extrusion direction as a function of the pole densities between the $\langle 11\bar{2}0 \rangle$ and $\langle 10\bar{1}0 \rangle$ poles for the Z-series alloys.

5. Discussion

ZK-series alloys

All extruded profiles of the ZK-series alloys showed inhomogeneous microstructures as a result of incomplete DRX. It is very likely that the presence of fine, intermetallic particles, developed during extrusion, are responsible for the retardation of the recrystallisation process. Alloys with a higher Zn content showed larger numbers of particles and also greater fractions of large, unrecrystallised grains, see Figs. 4.16-4.18. The development of the $\langle 10\bar{1}0 \rangle$ texture component in all the alloys of this series is associated with the presence of these large grains, see Fig. 4.26. This was confirmed by EBSD measurements on ZK40 alloys indirectly extruded at 1 [m/min] as shown in Fig. 5.8. Unrecrystallised grains show the orientation of the $\langle 10\bar{1}0 \rangle$ component, see Fig. 5.8 (a), whereas most of the recrystallised grains tend to orient preferentially towards the $\langle 11\bar{2}0 \rangle$ pole, see Fig. 5.8 (b). Others investigators have also correlated the $\langle 10\bar{1}0 \rangle$ component with the orientation of large, unrecrystallised grains in other extruded magnesium alloys [77, 86, 105].

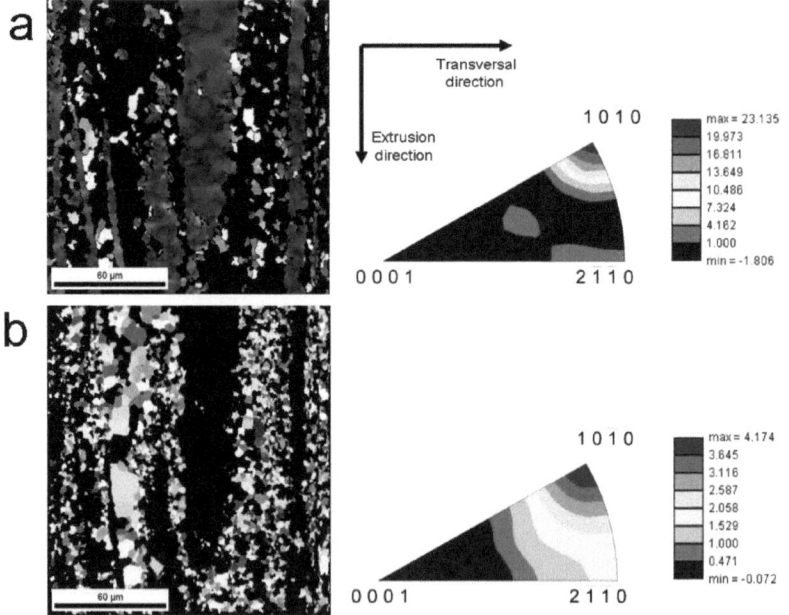

Fig. 5.8. EBSD map showing a sharp $\langle 10\bar{1}0 \rangle$ orientation of unrecrystallised regions (a) and the orientation of recrystallised regions (b) in extruded ZK40 alloy at 1 [m/min].

5. Discussion

The progress of DRX in the indirectly extruded ZK-series alloys can be followed by examining the intensity distributions shown in Fig. 5.9. Gradients of intensity from the lowest to the highest extrusion speed are shown as ΔI. These changes in texture can be correlated with the microstructures shown in Figs. 4.16-4.18. For example, the alloys extruded at 1 [m/min] have the greatest fractions of large, unrecrystallised grains and as a direct consequence, the highest intensities on the $\langle 10\bar{1}0 \rangle$ pole. In this category, the highest intensities are shown by the ZK30, ZK40 and ZK60 alloys. The ZK10 and ZK20 alloys show a significant decrease in the maximum intensity as a result of their lower Zn content, which means fewer precipitates and consequently fewer obstacles to the progress of recrystallisation.

At an extrusion speed of 10 [m/min], more of the large grains become recrystallised due to the increased temperature and a pronounced decrease in the intensity on the $\langle 10\bar{1}0 \rangle$ pole is observed. Alloys with fewer precipitates, i.e. ZK10 and ZK20 show the lowest intensities on the $\langle 10\bar{1}0 \rangle$ pole and also the highest gradients (ΔI) of -8 and -6 [m.r.d], indicative of more advanced recrystallisation. As with the Z-series alloys, it is very likely that the recrystallised grains develop a texture component on the $\langle 11\bar{2}0 \rangle$ pole, however these alloys show only a uniform distribution toward the $\langle 11\bar{2}0 \rangle$ pole, a sign of not effective DRX.

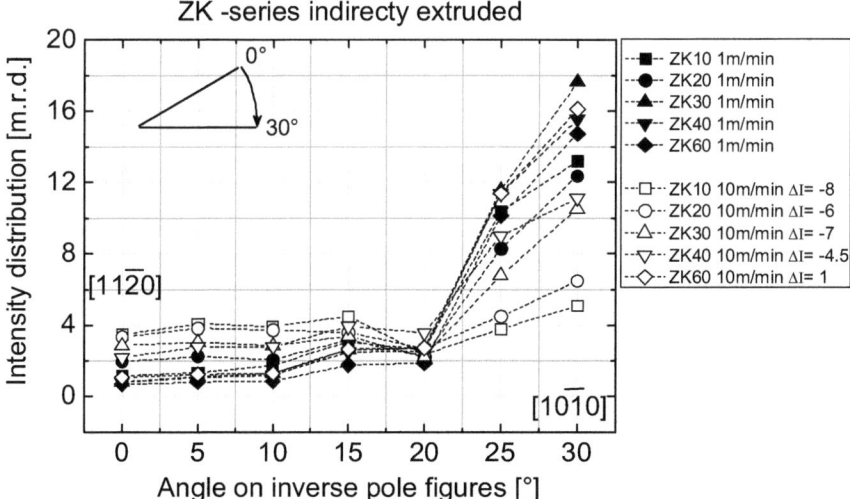

Fig. 5.9. Intensity distribution on the inverse pole figures in the extrusion direction as a function of the pole densities between the $\langle 11\bar{2}0 \rangle$ and $\langle 10\bar{1}0 \rangle$ poles for the ZK-series alloys correlated with their intensity gradient ΔI.

5. Discussion

Grain growth takes place at higher extrusion speeds and these coarser grains coexist with the large unrecrystallised grains. Considering that grain growth cannot occur before recrystallisation has finished, this could indicate that the temperature achieved during extrusion (~440 °C) was high enough to enable grain growth of the primary recrystallised grains, but not high enough to initiate recrystallisation in the large grains. Segregation of particles also gives areas free of particles where grain growth can be easily promoted.

In terms of deformation, the ZK-series alloys show lower Z values, therefore, higher temperatures and stresses are necessary for the effective activation of DRX. None of these alloys showed such effectiveness, which is reflected on inhomogeneous microstructures and weak $\langle 11\bar{2}0 \rangle$ textures components. Higher temperatures or higher stresses seem to be necessary to achieve effective DRX in these alloys. Higher extrusion temperatures cannot be used because ZK-alloys with higher Zn contents (as commercially applied) are at the limit of the solidus line of the binary Mg-Zn phase diagram. All the alloys studied in this work showed acceptable quality surfaces, only the ZK60 alloy extruded at 10 [m/min] showed traces of superficial hot-cracks, a sign of its limited extrudability. Other deformation processes involving higher stresses, such as hydrostatic extrusion would be more appropriate to achieve adequate DRX in these alloys and better control of the resulting grain sizes.

ZEK-series alloys

With the addition of Ce-Mischmetall even more grain refined microstructures were achieved after extrusion. This has been attributed to the higher fraction of precipitates present in the as-cast condition, which enhanced the Zener pinning effect during extrusion. Thus recrystallisation should also be hindered, similar to the ZK-series. The result of this is an increased fraction of large unrecrystallised grains. Nevertheless with the increase in the profile temperature (~450 °C), precipitates do not seem to hinder the recrystallisation progress anymore. The number of large grains was considerably reduced and grain growth was more homogeneous than that observed in the ZK-series.

Fig. 5.10 shows the intensity variation for the indirectly extruded ZEK-series alloys. The microstructures shown in Figs. 4.21 and 4.22 can be correlated with the texture changes. The higher portion of large, unrecrystallised grains is again correlated very well with even higher intensities on the $\langle 10\bar{1}0 \rangle$ pole. Similar to the ZK-series alloys extruded at higher speed, the

5. Discussion

maximum intensity was decreased. However, considering that these alloys showed the highest intensities, the gradients were most significant (between 10 and 13.5 [m.r.d], see ΔI gradients). The microstructures match with this change; most of the large grains have recrystallised, although some are still observed especially in the alloys with higher Zn contents. Recrystallisation has progressed considerably, which is also followed by grain growth. This grain growth is better controlled by the uniformly distributed, fine particles.

If the gradients of intensity are compared between the ZK and the ZEK-series alloys, it is clear that the recrystallisation process is faster in alloys with the Ce-MM addition, although DRX is fastest in the pure Mg and the binary Z-series alloys.

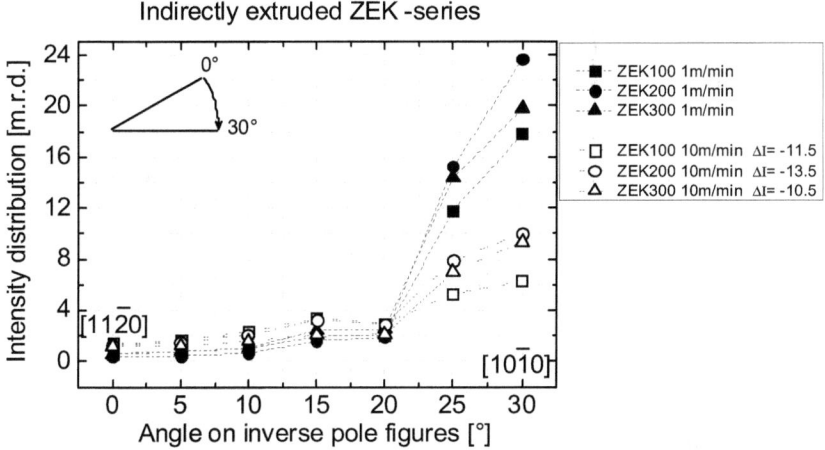

Fig. 5.10. Intensity distribution on the inverse pole figures in the extrusion direction as a function of the pole densities between the $\langle 11\bar{2}0 \rangle$ and $\langle 10\bar{1}0 \rangle$ poles for the ZEK –series alloys correlated with their intensity gradient ΔI.

It is important to note that this decrease in the maximum intensity is not accompanied by an increase in the distribution toward the $\langle 11\bar{2}0 \rangle$ pole, in contrast with the behaviour in the Z and ZK-series alloys. Instead, the maximum intensity tends to orient toward the $\langle 11\bar{2}1 \rangle$ pole, see Fig. 4.27. An EBSD measurement on the ZEK300 alloy extruded at 1 [m/min] shows that this texture component corresponds only to these new recrystallised grains and represents an area fraction of 0.06 (see Fig. 5.11). This component tended to increase after the annealing treatment and will be discussed in the annealing treatment section.

5. Discussion

The addition of Zn does not have a significant effect on the variation of the intensities, although the alloys with higher Zn contents showed slightly higher intensities, likely due to a slight increase in the precipitate volume fractions.

In terms of deformation response, the Ce-MM containing alloys tended to have considerably increased Z values, thus higher stresses and temperatures are necessary to activate DRX. However, these alloys are able to withstand temperatures of 450°C without any appearance of hot-cracking. This enables the temperature necessary for effective DRX to be reached with a simultaneous fall in the flow stresses, which is characteristic of an effective DRX process. In this respect it can be concluded that for similar extrusion conditions the kinetics of DRX are faster than in the ZK-series alloys.

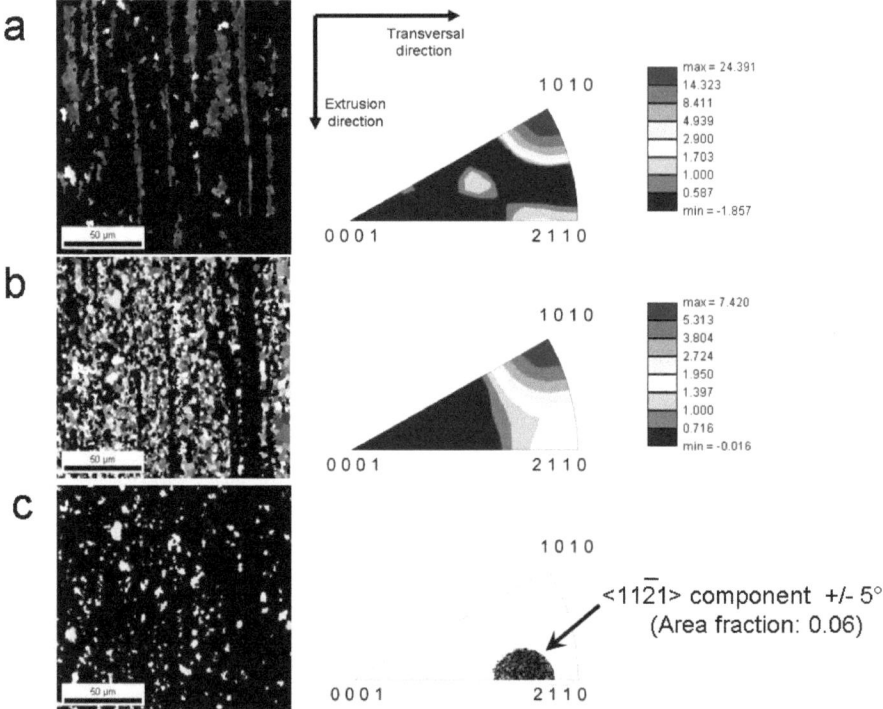

Fig. 5.11. EBSD map showing a sharp $\langle 10\bar{1}0 \rangle$ orientation of unrecrystallised regions (a) and the orientation of recrystallised regions (b) in extruded ZEK300 alloy at 1 [m/min]. New recrystallsed grains oriented on $\langle 11\bar{2}1 \rangle$ texture component are shown in (c).

5. Discussion

5.3.1 Recrystallisation as result of the annealing treatment (Static RX)

After the annealing treatment it was possible to observe some advance in the recrystallisation process, especially in those alloys with inhomogeneous microstructures. The changes shown in section 4.6 are a consequence of static recrystallisation (RX), where the deformation is absent and the driving force for recrystallisation depends on the release of the stored energy of the deformed, unrecrystallised grains. Basically, the ZK and ZEK-series alloys recrystallised statically as a function of the fraction of unrecrystallised large grains present in the microstructure of extruded alloys. Nonetheless, each group of alloys showed some particularities that deserve attention.

ZK-series alloys

The main effects of the annealing treatment on the distribution of intensities for the ZK-series alloys are shown in Figs. 5.11 a) and b). Alloys with a higher fraction of unrecrystallised grains, i.e. alloys extruded at 1 [m/min] and the ZK60 alloy extruded at 10 [m/min], have undergone considerable SRX as revealed by the higher gradients of intensity after the annealing treatment, see gradients ΔI in Fig. 5.11. It is interesting to note that the alloys extruded at 1 [m/min] showed similar gradients. This behaviour is opposite to that observed for DRX as a consequence of extrusion, where the addition of Zn or a larger fraction of precipitates reduced the resulting gradients, see Fig. 5.9. This point is very important because it confirms that static RX is controlled principally by the fraction of unrecrystallised grains present after extrusion.

Another important result is shown by the ZK20 alloy after the annealing treatment. In this alloy, the $\langle 11\bar{2}0 \rangle$ component has appeared with a clear tendency to continue increasing, similar to the behaviour shown by the binary Z-series alloys. Therefore the $\langle 11\bar{2}0 \rangle$ component is definitely the preferred orientation of the recrystallised grains in the Z and ZK-series alloys. The sharpening of the $\langle 11\bar{2}0 \rangle$ component indicates that recrystallisation is carried out by bulging and rotational DRX mechanism [5, 17, 79], which was unaccomplished dynamically.

5. Discussion

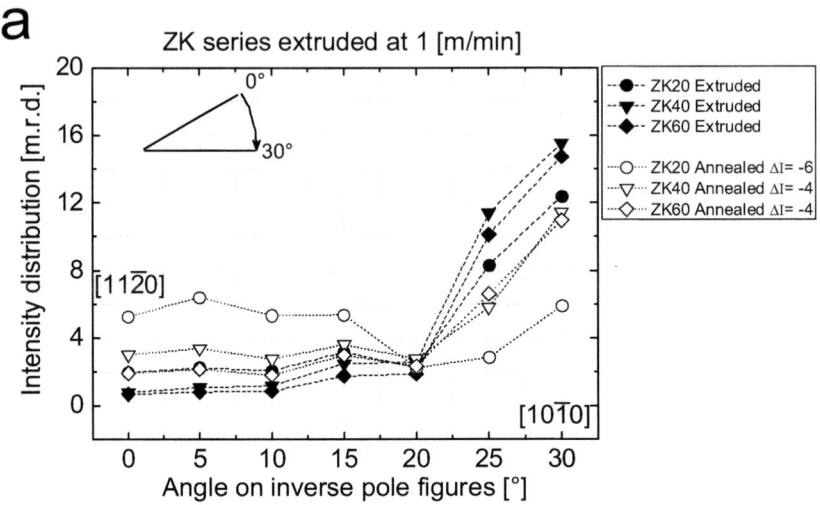

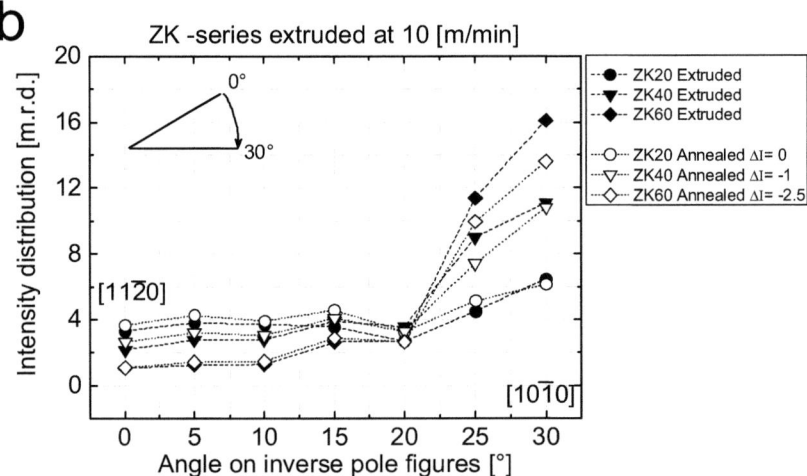

Fig. 5.12. Intensity distribution on the inverse pole figures in the extrusion direction as a function of the pole densities between the $\langle 11\bar{2}0 \rangle$ and $\langle 10\bar{1}0 \rangle$ poles for the ZK-series alloys indirectly extruded at a) 1 [m/min] and b) 10 [m/min], before and after annealing at 400 °C for 1 hour. Intensity gradients (ΔI) are shown.

5. Discussion

Recrystallisation of the ZEK-series after annealing treatment

Similar to the ZK-series, the ZEK-series alloys with higher fractions of unrecrystallised grains in the extruded condition (i.e. extruded at 1 [m/min]) showed the most significant decreases in intensity on the $\langle 10\bar{1}0 \rangle$ pole after the annealing treatment, see Fig. 5.12 a). Although these gradients are lower than those resulting from the extrusion process (compare Figs. 5.12 a) and 5.10), they are also uniform with a slight tendency to increase with the Zn addition.

On the other hand, alloys with lower fractions of unrecrystallised grains, i.e. those extruded at 10 [m/min], did not show apparently any change in intensity distribution after annealing (compare Figs. 5.12 b and Fig. 5.10), although their microstructures showed the presence of several new recrystallised grains and the inverse pole figures showed an increase in intensity of the $\langle 11\bar{2}1 \rangle$ component, see Fig. 4.41. This component also developed as a result of the extrusion process, but with a very weak intensity, (see Figs. 4.27 and 4.28). Thus, the presence of the $\langle 11\bar{2}1 \rangle$ component is attributed to the addition of Ce-MM and a recrystallisation mechanism different to that of bulging.

Given that the Ce-Mischmetall used in this work consists mainly of Ce (57.4 wt. %) and La (27.3 wt. %; see Table 3.1), it would seem that the development of the $\langle 11\bar{2}1 \rangle$ texture component is directly associated with the presence of these rare earth elements. In the literature, the formation of this texture component has been attributed exclusively to the role of rare earth additions, in particular to additions of La and Gd [90]. However, in the present work, weaker but well defined $\langle 11\bar{2}1 \rangle$ components were also found in the ZK-series alloys with no rare earth additions, (see Figs. 4.26, 4.36 and 4.37). Moreover, extruded Mg-Mn alloys with additions of Ce, Y, and Nd, also developed this component [88]. This suggests that other factors can influence the development of this $\langle 11\bar{2}1 \rangle$ component and that its presence cannot be attributed solely to the effects of rare earth additions.

The ZEK-series alloys extruded at 10 [m/min] did not show any change with respect to the intensity of the $\langle 10\bar{1}0 \rangle$ pole after annealing, see Fig. 5.10. The reason for this is that most of the large grains have already recrystallised after extrusion. This is confirmed by the low intensities on the $\langle 10\bar{1}0 \rangle$ pole. In other words, most of the stored energy of the deformed grains necessary to induce static recrystallisation is no longer available. However, recrystallisation has been observed and the intensity of the $\langle 11\bar{2}1 \rangle$ component has increased. This increase is likely to occur at the expense of the distributed intensity between the $\langle 10\bar{1}0 \rangle$

5. Discussion

pole and the $\langle 11\bar{2}1 \rangle$ component already present in samples in the extruded condition; see Fig. 4.27.

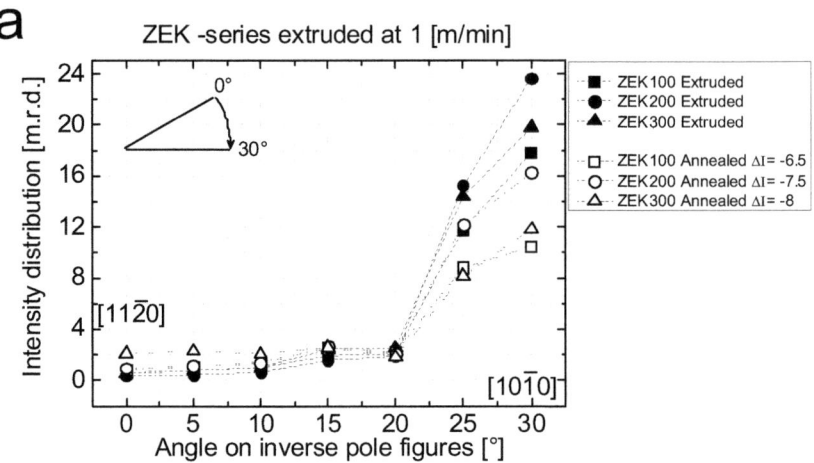

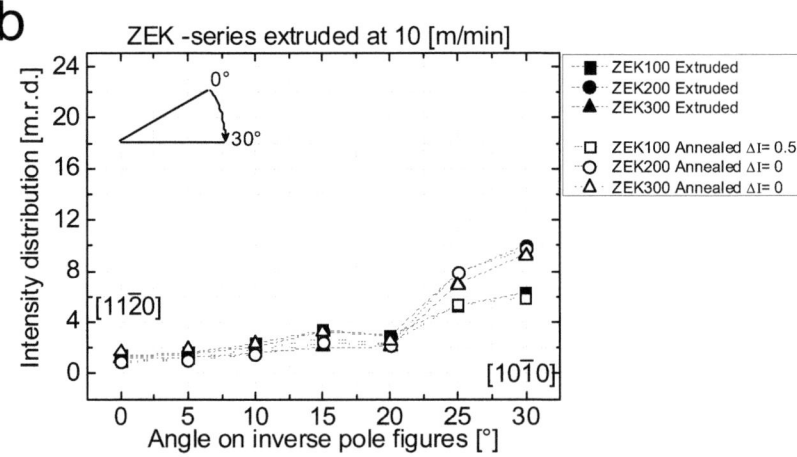

Fig. 5.13. Intensity distribution on the inverse pole figures in the extrusion direction as a function of the pole densities between the $\langle 11\bar{2}0 \rangle$ and $\langle 10\bar{1}0 \rangle$ poles for the ZEK-series alloys indirectly extruded at a) 1 [m/min] and b) 10 [m/min], before and after annealing at 400 °C for 1 hour. Intensity gradients (ΔI) are shown.

5.4 Yield strength, yield strength asymmetry and the Hall-Petch relationship

The grain size and texture have a strong influence on important mechanical properties of wrought magnesium alloys such as the yield strength, the yield asymmetry and strain to fracture. This contrasts significantly with the behaviour of cast alloys, where the yield asymmetry is negligible [88]. This section is focussed on such mechanical values and Hall-Petch plots are used to correlate the microstructure and yield strength of the extruded profiles.

Until now it has been considered that the yield strengths obtained are in good agreement with an expected increase as a result of grain refinement. However, at this point there is no possibility to distinguish quantitatively between the effect of the average grain size resulting from processing and effects due to the alloying element additions. Hall-Petch plots have been used successfully to separate these effects in extruded magnesium alloys [3, 88, 96, 118].

It is very important to point out that the materials used in this study do not allow a "true" Hall-Petch analysis, in the sense that the parameters σ_0 and k are determined, due to the use of textured material. The fact that different yield strengths are obtained in tension and compression is an indicator for this textured condition. However, the yield asymmetry behaviour is also influenced by the texture components and their intensities. This allows the integration of the texture information in the plots to understand better its influence on the yield behaviour.

Pure Mg and Z-series alloys

Fig. 5.14 shows the Hall-Petch plot for this group of alloys In order to avoid incongruous interpretation of the plot; the tensile yield strength values of the samples showing micro-yielding phenomena were not included. The values shown in the plot correspond only to the TYS values of the Z-series alloys indirectly extruded at 1 [m/min] and the CYS values of the whole group.

In order to extend the limited data available, in particular with respect to the range of grain sizes, the yield strength data for the same materials extruded hydrostatically have been included in this analysis. Hydrostatic extrusion has been used to produce grain-refined microstructures [99]. The TYS and CYS values of the hydrostatically extruded materials have been added to Fig. 5.14 (indicated in gray).

5. Discussion

Prior to an analysis of the yield strength, it is important to comment on the unexpected micro-yielding phenomena shown by this group of alloys. Samples showing micro-yielding and those free of this can be classified according to microstructural aspects. The micro-yielding group possess a sharp and well defined $\langle 11\bar{2}0 \rangle$ texture component, whereas the other group showed a tilted $\langle 11\bar{2}0 \rangle$ component which is generally accompanied by a $\langle 10\bar{1}0 \rangle$ component, see Fig. 4.25 and Table 4.1. There is another common characteristic in the micro-yielding group; they possess very coarse grain sizes (between 45 and 80 µm). In contrast, the other group showed microstructures with finer grain sizes (between 10 and 40 µm). Thus, micro-yielding phenomena are associated with the development of the $\langle 11\bar{2}0 \rangle$ texture component combined with a coarse grain size.

Returning to the Hall-Petch plot, the tensile and compressive yield strength data for the Z-series alloys have been connected with straight lines. They show basically typical Hall-Petch behaviour, where the yield strength is increased thanks to the grain refinement achieved in the hydrostatically extruded alloys. The progressive Zn addition did not produce any change in the yield strength. However, in comparison with pure Mg it is demonstrated that a significant improvement can be attributed to the Zn addition. Continuing with the comparison with pure Mg, it is clear that the pronounced yield asymmetry can also be attributed to the Zn addition. The fact that even grain-refined microstructures showed a pronounced yield asymmetry, reinforces this point.

In spite of the above mentioned differences in the textures of this group, in general they are very similar with respect to the orientation of the basal planes which lie parallel to the extrusion direction. Therefore, $\{10\bar{1}2\}$ twins are more easily activated in compression than in tension, as reflected in the flow curves (see Fig. 4.28). Considering that between pure Mg and the Z-series alloys there is no considerable difference in the texture intensities and components and no significant difference in microstructure and grain size, then the substantial differences in yield asymmetry can be only understood as a solute element effect. Additions of Zn solute in magnesium single crystals produce a hardening effect on the basal planes and a softening effect on the prismatic and pyramidal planes [11, 12]. The hardening of the basal planes strongly affects the initiation of plastic deformation during tensile testing, because plastic deformation in this direction depends mainly on basal slip. The result of this is a considerable rise in the TYS with respect to that of pure Mg and a marked yield asymmetry.

5. Discussion

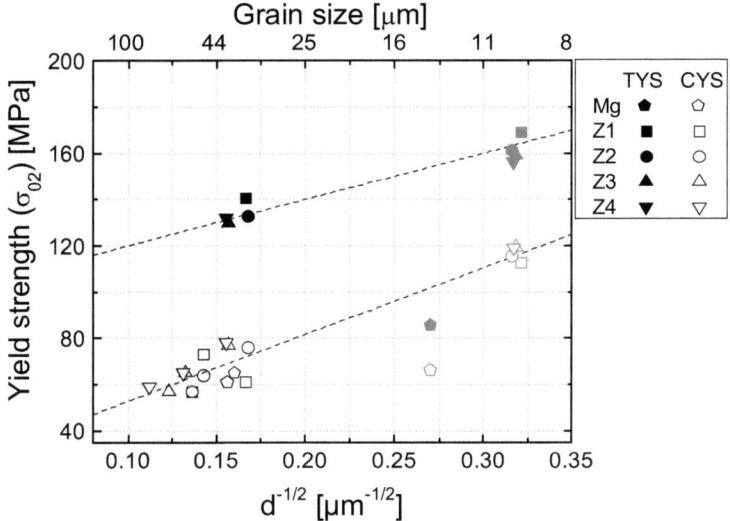

Fig. 5.14. Tensile and compressive yield strength *vs.* inverse square root of average grain size for indirectly extruded pure Mg and Z-series alloys. CYS and TYS of grain refined pure Mg and Z-series alloys hydrostatically extruded has been also added to this plot (gray colour).

On other hand, yield asymmetry of the indirectly extruded alloys is slightly greater than that of the hydrostatically extruded alloys; however this is only an effect of the grain size as the plot itself indicates. The activation of twinning in coarser grains is easier, because there are less grain boundaries acting as barriers for twinning. The start of plastic deformation in compression is quite dependent on the activation of twinning. Thus with coarser grains, the CYS is strongly reduced and the yield asymmetry accentuated.

As mentioned above, the addition of Zn solute to Mg is accompanied by a softening of the prismatic and pyramidal planes during deformation at room temperature. This ability of Zn as solute seems to be well rounded during tensile test of these alloys, where elongation of these alloys was significantly improved with the progressive Zn addition, see Fig. 4.29 d). Actually, this progressive improvement of elongation was observed only in the Z-series alloys. Higher elongations are reached only during tensile test, because twinning is more difficult to activate increasing with this the stress necessary to overcome yield strength limit. Tensile yield stresses are higher than those of compression and enough to activate non-basal slip. Thus, prismatic and pyramidal slip is feasible to activate after yielding during tensile test. Since Zn

5. Discussion

addition reduces progressively the CRSS of such slip systems, then elongation should increases as a concomitant effect. Oppose to this, profusely activation of twinning is carried out during compression test reducing the yield strength and limiting the changes to activate non-basal slip. As the deformation proceeds, work hardening is prone to become facilitated by the interaction between twins and dislocations that subsequently form, increasing the ultimate strength UCS (compare UTS and UCS Fig. 4.29 c)) at cost of lower elongations, (see Table 4.2).

ZK-series alloys

With the addition of Zr to the binary Z series, a significant increase in the yield strength with decreasing grain size is observed; see Fig. 5.15. The guide lines representing yield strength shows that the slope in compression is steeper than in tension. This behaviour is in concordance with that observed in the binary alloys and can be explained by the fact that $\{10\bar{1}2\}$ twinning is geometrically more favoured in compression than in tension and that the activity of twinning decreases with decreasing grain size [3, 96]. Therefore, with the considerable decrease in the grain size, twinning is more difficult to activate. Thus, the CYS increases and the yield asymmetry is simultaneously reduced.

All extruded ZK-series alloys developed a common $\langle 10\bar{1}0 \rangle$ texture component with differences only in the maximum intensity levels. The presence of the $\langle 10\bar{1}0 \rangle$ component did not produce any change with respect to the orientation of the basal planes which lie parallel to the extrusion direction. Thus, the texture change in this regard is not considered as the reason for the significant decrease in the yield asymmetry. As previously discussed, changes in the intensity levels have been correlated mainly with the recrystallisation process. In resume, changes in intensity level between the $\langle 10\bar{1}0 \rangle$ and $\langle 11\bar{2}0 \rangle$ prismatic components do not influence the yield strength or yield asymmetry at all. However, the high intensities registered by texture measurements on these alloys, are related to the hardening behaviour observed in the flow curves during tensile testing, see Fig. 4.30 a). Instead of the parabolic behaviour observed in the binary Z-series alloys, ZK-series alloys extruded at 1 [m/min] showed semi-linear behaviour. This behaviour is directly related to the inhomogeneous microstructure. The microstructure consists of large unrecrystallised grains together with new recrystallised grains, but the high intensities indicate that the former are dominant. Therefore, during plastic deformation, whereas the dominant large grains tend to develop twins, the new grains tend to

5. Discussion

deform by slip. Thus inhomogeneous deformation without increasing hardening would take place.

With increasing extrusion speed, a greater fraction of large deformed grains have already recrystallised dynamically and then the microstructure consists mostly of recrystallized grains with similar size. Twinning and twin size are now restricted by the new recrystallised grain size, leading to dislocation controlled slip deformation. As a result, the flow curves show a parabolic slip-dominated deformation, see Fig. 4.30 (c).

One important contribution of the Hall-Petch plot is the possibility to observe changes related to the chemical composition. The difference between the ZK-series alloys is only an increasing content of Zn. The effect of Zn on the yield strength can be clearly observed as a vertical increase proportional to the Zn addition if we compare the yield strength for the alloys with a similar grain size of ~25 µm. It is clear that the Zn addition contributes to increase the yield strength. However, it is possible that the presence of intermetallic particles, instead of the content of Zn solute, is responsible for this strengthening. This contrast the behaviour observed in the binary alloys, where no difference in yield strength is observed between alloys with different Zn contents.

In a word, it can be stated that grain refinement is the main factor responsible for a decrease of the yield asymmetry at least for the alloys studied in the present work. This is confirmed by work on a hydrostatically extruded ZK30 alloy, where finer grain sizes contributed to eliminate completely the yield asymmetry [96].

5. Discussion

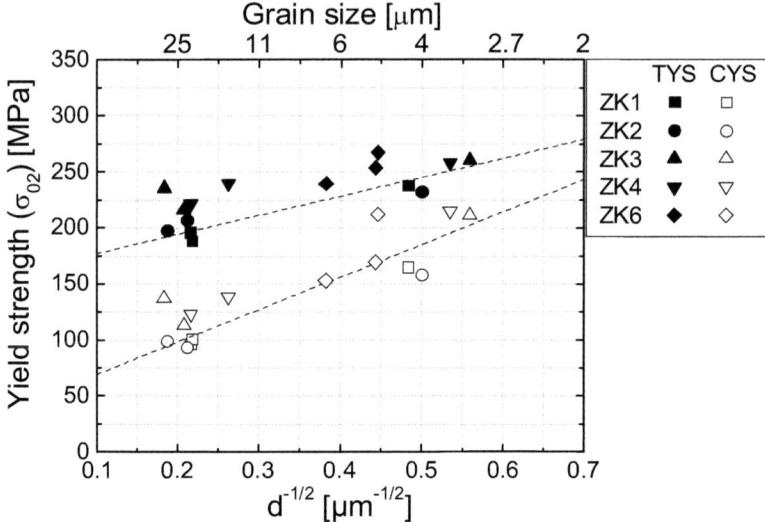

Fig. 5.15. Tensile and compressive yield strength *vs.* inverse square root of average grain size for indirectly extruded ZK-series alloys.

ZEK-series alloys

With the addition of Ce-MM to the ZK-series, the resulting microstructures were the finest ones of the present work leading to the highest levels of the tensile and compressive yield strengths, see Fig. 5.16. Within the scatter of data no significant difference is visible as a function of the Zn content. In spite of the fine grain sizes achieved for this group of alloys they do not show any effect on the yield asymmetry. However, alloys with coarser grains show a reduced yield asymmetry. This irregular behaviour has its origin in the texture and inhomogeneous microstructure of this group of alloys.

First, following the behaviour observed for the ZK-series alloys, the reasons for an increased asymmetry in the most grain refined alloys (extruded at 1 [m/min]) will be explained. The textures of this group of alloys indicated that the $\langle 10\overline{1}0 \rangle$ component is dominant, as in the ZK –series alloys. However, the intensities are significantly higher particularly in the most grain-refined alloys. In other words, the presence of the large unrecrystallised grains is even more dominant than that observed in the grain refined ZK-series. Therefore, plastic deformation of these alloys depends mostly on the large unrecrystallised grains, where twinning is easily activated, leading to an increase in the yield asymmetry. New recrystallised grains do not

5. Discussion

dominate, but their fraction is represented by the finest grains of the alloys studied. They are responsible for enhancing the yield strength.

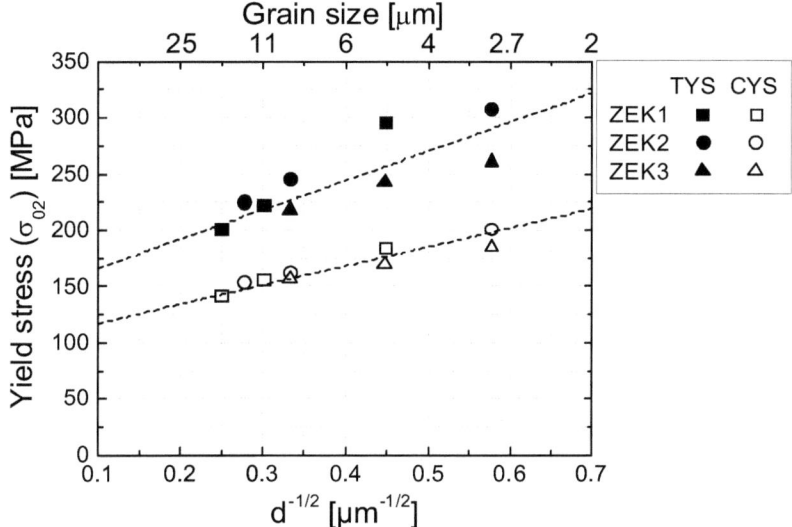

Fig. 5.16. Tensile and compressive yield strength *vs.* inverse square root of average grain size for indirectly and hydrostatic extruded ZEK – series alloys. I_{max}: Maximum intensity in [MRD].

This microstructure also led to an even more inhomogeneous deformation during tensile testing, see Fig. 4.32. The flow curves of these alloys showed a non-hardening behaviour during plastic deformation.

On other hand, alloys extruded at 10 [m/min] showed a significant decrease in the maximum texture intensity, which has been associated with the considerable progress of the DRX process, together with limited grain growth and more homogenous microstructures. The fact that the microstructure becomes homogenous is also reflected during tensile testing, where the flow curves show a parabolic slip-controlled behaviour (see Fig. 4.32 (c)). The reduced yield asymmetry shown by these alloys can be explained by a larger fraction or new dynamically recrystallised grains, whose growth is strongly inhibited by the Zener-pinning mechanism. Hence, a more homogenous distribution of fine recrystallised grains obtained at higher extrusion speeds restricts twinning and reduces the yield asymmetry.

5. Discussion

In addition, the development of the $\langle 11\bar{2}1 \rangle$ texture component could also contribute somewhat. This component tends to continue developing after the annealing treatment and probably this effect can be enhanced with the Ce-MM addition. A considerable reduction in the yield asymmetry has been observed in extruded magnesium alloys showing this component [88].

6. Conclusions

In the present work the effect of the addition of zinc, zirconium and Ce-Mischmetal on the casting, indirect extrusion processing, microstructural development and resulting mechanical properties of magnesium and magnesium alloys were investigated. In each link of this experimental chain, specific aims corresponding to those described in the introduction to this work were investigated. From the experimental results and their analysis, the following conclusions can be drawn:

Cast alloys and their influence on indirect extrusion processing

Grain refinement was a generalised effect of the addition of alloying elements on the microstructures of the cast magnesium billets used in this work. Zr, the most powerful grain refiner, produced as expected the most significant reductions in grain size, although Zn and Ce-MM additions also contributed to grain refinement. Zn was the only element whose content was progressively changed in the different alloys used in this work. The binary Mg-Zn alloys allowed an analysis of the grain refining mechanism. This was explained on the basis of a thermodynamically based grain growth factor (Q), which can be determined from the binary phase diagrams of eutectic systems. The grain refinement mechanism is controlled by the Zn solute content and is associated with constitutional undercooling of the melt. The importance of the Q factor is related to the possibility of predicting the grain size of the cast billet, a parameter which influenced strongly the deformation response of the alloy during extrusion. The Q mechanism could also be applied to the multi-component alloys such as those of the ZK and ZEK-series studied in this work; nevertheless it is necessary to consider other variables in order to adapt this model, e.g. to the particular inoculation mechanism of Zr, where both dissolved Zr and undissolved Zr particles participate and a peritectic reaction takes place during solidification.

Grain size and chemical composition were the principal variables of the billets used for the indirect extrusion experiments. Their influence on processing was analysed in terms of the extrusion forces and temperatures measured locally during the process. The initial grain size of the cast and homogenised billets showed a linear relationship with the profile temperature, having a specific slope for each extrusion speed. This relationship was negligibly affected by alloy composition and allowed the determination of the resulting temperature as a consequence of deformation for a specific initial grain size.

6. Conclusions

The steady state forces registered during extrusion increased also for grain refined cast billets. Still, these parameters did not show a clear correlation between them. Steady state force is a value dependent on the dynamical deformation response of each alloy and it was better correlated with the final microstructure than with the initial one. A noteworthy relationship was found between steady state forces and profile temperatures, showing a linear relationship specific for a constant extrusion speed. Correlations between *initial grain size–temperature* and *temperature-steady state force* allow us to predict the resulting extrusion temperature and load as a function of the initial grain size for a specific alloy.

Deformation response of the alloys during extrusion processing

The deformation response during extrusion of the selected studied alloys was analysed in terms of the deformation temperature (profile temperature), strain rate ($\dot{\varepsilon}_t$) and strain; i.e. mean flow stress (σ_f) determined from the extrusion force in a stationary state. The Zener-Hollomon parameter (Z) was used to correlate these variables. The correlation between flow stress and Z values showed that the deformation response is alloy composition dependent. High Z values involve low flow stresses and small increases in temperature above 300°C and low Z values the opposite. The most important contribution of the determination of Z values is the possibility to correlate alloy dependent deformation variables during extrusion with the resulting microstructure, i.e. the resulting recrystallised grain size and texture.

Microstructural development as result of extrusion processing

The average, recrystallised grain sizes of the extruded alloys are related to the Z values. The larger the parameter Z is, the finer the recrystallised grain size. This relationship allows us to predict the resulting grain size as a function of temperature, strain and strain rate. However, the fraction of recrystallised grains and the extend of subsequent grain growth are alloy dependent, because the alloy composition determines the mechanism and kinetics of recrystallisation.

Texture measurements performed on extruded profiles provided additional information on the recrystallisation process in each group of alloys. They revealed a particular development of textures components in function of alloys composition and Z parameter. This development was also correlated with the homogeneity of the recrystallised microstructure and the roll of intermetallic particles. Annealing treatment of selected extruded profiles revealed a subsequent and singular progress of the static recrystallisation process

6. Conclusions

Mechanical properties of the extruded profiles

Hall-Petch plots were used to separate the effects of grain size from that of the alloying elements on the yield strength and the yield asymmetry. Independent of the texture changes described, it is clear that the basal planes in all alloys lie parallel to the extrusion direction. Thus, yielding of the alloys was basically dependent on twinning and controlled by grain size, in particular during compression testing. With a reduction in grain size, tensile twins were inhibited during compression testing, so that the compressive yield strength increased and simultaneously the resulting yield asymmetry was reduced.

With the addition of Zn to pure Mg, the yield strength was significantly improved; however, the yield asymmetry was increased as consequence of a hardening of the basal slip by the Zn addition. Elongations achieved during tensile testing of these alloys were considerably improved with the Zn addition. This is attributing to softening of the prismatic and pyramidal slip attained with the Zn addition. With the Zr addition to the binary alloys, a significant increase in the yield strength with decreasing grain size was obtained. This improvement was also attributed to the presence of intermetallic precipitates. The yield asymmetry was also significantly reduced by grain refinement and it has been shown that the yield asymmetry can be eliminated by enhanced grain refinement of these alloys.

With the Ce-MM addition to the ZK-series alloys, the resulting microstructures were the finest ones of the present work leading to the highest tensile and compressive yield strengths. In contrast to the other alloys, a reduced yield asymmetry was observed for these alloys. This was explained on the basis of a faster DRX process attained at high temperatures, which changed the microstructure from inhomogeneous to homogeneous. In addition, grain growth was significantly inhibited by the considerable presence of complex intermetallic particles homogenously distributed in the matrix. The development of the $\langle 11\bar{2}1 \rangle$ component shown by these alloys could also contribute to reduce yield asymmetry.

The experimental sequence followed in this work and the resulting correlations between initial alloy conditions and process variables indicate that the resulting microstructure and mechanical properties can be estimated by an appropriate set-up of selected alloy compositions and process parameters. Most of these correlations were based on proved phenomenological assumptions, which make them reliable values of reference for additional research and development of appropriate magnesium alloys for such processing.

7. Bibliography

[1] H. Friedrich, S. Schumann, Research for a "new age of magnesium" in the automotive industry, Journal of Materials Processing Technology 117 (2001) 276-281.
[2] B. Closset, Magnesium: Present and Future Opportunities, Proceedings of the 2nd International Light Metals Technology Conference 2005 (2005) 9-13.
[3] J. Bohlen, P. Dobron, J. Swiostek, D. Letzig, F. Chmelik, P. Lukac, K.U. Kainer, On the influence of the grain size and solute content on the AE response of magnesium alloys tested in tension and compression, Materials Science and Engineering: A 462 (2007) 302-306.
[4] D.H. StJohn, M. Qian, M.A. Easton, P. Cao, Z. Hildebrand, Grain refinement of magnesium alloys, Metallurgical and Materials Transactions A 36A (2005) 1669-1679.
[5] A. Galiyev, R. Kaibyshev, G. Gottstein, Correlation of plastic deformation and dynamic recrystallization in magnesium alloy ZK60, Acta Materialia 49 (2001) 1199-1207.
[6] M.R. Barnett, Recrystallization during and following hot working of magnesium alloy AZ31, Materials Science Forum 419-422 (2003) 503-508.
[7] E.A. Ball, P.B. Prangnell, Tensile-compressive yield asymmetries in high strength wrought magnesium, Scripta Metallurgica et Materialia 31 (1994) 111-116.
[8] J. Bohlen, M.R. Nurnberg, J.W. Senn, D. Letzig, S.R. Agnew, The texture and anisotropy of magnesium-zinc-rare earth alloy sheets, Acta Materialia 55 (2007) 2101-2112.
[9] J.P. Doan, G. Ansel, Some effects of zirconium on extrusion properties of magnesium-base alloys containing zinc, AIME (1947) 286-305.
[10] C.J. Bettles, M.A. Gibson, Current Wrought Magnesium Alloys: Strengths and Weaknesses, JOM (The Journal of the Minerals, Metals & Materials Society) (2005) 46-49.
[11] A. Akhtar, E. Teghtsoonian, Solid solution strengthening of magnesium single crystals - I. Alloying behaviuor in basal slip, Acta Metallurgica (1969) 1339-1349.
[12] A. Akhtar, E. Teghtsoonian, Solid solution strengthening of magnesium single crystals - II. The effect of solute on the ease of prismatic slip, Acta Metallurgica (1969) 1351-1356.
[13] C.H. Cáceres, A.H. Blake, Solute and temperature effects on the strain hardening behavior of Mg-Zn solid solutions, Materials Science Forum 567-568 (2008) 45-50.
[14] C.H. Cáceres, A.H. Blake, The Strength of Concentrated Mg-Zn Solid Solutions, physica status solidi (a) 194 (2002) 147-158.
[15] T.E. Leontis, The properties of sand cast magnesium-rare earth alloys, Journal of metals (AIME) 185 (1949) 968-983.
[16] Magnesium Taschenbuch, First ed., Aluminium - Verlag, Düsseldorf, Germany, 2000.
[17] S.E. Ion, F. Humphreys, S.H. White, Dynamic recrystallisation and the development of microstructure during the high temperature deformation of magnesium, Acta Metallurgica 30 (1982) 1909-1919.
[18] T. Obara, H. Yoshinga, S. Morozumi, {1122}{1123} Slip system in magnesium, Acta Metallurgica 21 (1973) 845-853.
[19] P.G. Patridge, The crystallography and deformation modes of hexagonal close-packed metals, Metallurgical Reviews 12 (1967) 169-194.
[20] R.V. Mises, Mechanik der plastischen Formänderung von Kristallen, Zeitschrift für Angewandte Mathematik und Mechanik 8 (1928) 161-185.
[21] B.D. Cullity, S.R. Stock, Elements of X-Ray Diffraction, Third ed., Prentice Hall, 2001.

7. Bibliography

[22] G.E. Dieter, Mechanical Metallurgy, SI Metric ed., McGraw-Hill Co., London, UK, 1988.
[23] D.W. Brown, S. Agnew, M.A.M. Bourke, T.M. Holden, S.C. Vogel, Internal strain and texture evolution during deformation twinning in magnesium Materials Science and Engineering A 399 (2005) 1-12.
[24] J. Koike, Enhanced deformation mechanisms by anisotropic plasticity in polycrstalline Mg alloys at room temperature, Metallurgical and Materials Transactions A 36A (2005) 1689-1696.
[25] S.R. Agnew, M.H. Yoo, C.N. Tome, Application of texture simulation to understanding mechanical behavior of Mg and solid solution alloys containing Li or Y, Acta Materialia 49 (2001) 4277-4289.
[26] S. Agnew, Plastic anisotropy of magnesium alloy AZ31B sheet, Magnesium Technology 2002 (2002).
[27] B.C. Wonsiewicz, W.A. Backofen, Plasticity of magnesium crystals, Transactions of the metallurgical society of AIME 239 (1967) 1422-1431.
[28] M. Yoo, Slip, twinning and fracture in hexagonal close-packet metals, Metallurgical Transactions A 12A (1981) 409-418.
[29] J.W. Christian, S. Mahajan, Deformation twinning, Progress in Materials Science 39 (1995) 1-157.
[30] M.R. Barnett, Twinning and the ductility of magnesium alloys: Part I: "Tension" twins, Materials Science and Engineering: A 464 (2007) 1-7.
[31] M.R. Barnett, Z. Keshavarz, A.G. Beer, D. Atwell, Influence of grain size on the compressive deformation of wrought Mg-3Al-1Zn, Acta Materialia 52 (2004) 5093-5103.
[32] Z. Keshavarz, M.R. Barnett, In-situ investigation of twinning behaviour in Mg-3Al-1Zn, Magnesium Technology 2005 (2005) 171-175.
[33] Y.N. Wang, J.C. Huang, The role of twinning and untwinning in yielding behavior in hot-extruded Mg-Al-Zn alloy, Acta Materialia 55 (2007) 897-905.
[34] Magnesium and Magnesium Alloys, ASM International, 1999.
[35] E.F. Emley, Principles of Magnesium Technology, Pergamon Press, London, 1966.
[36] Y.C. Lee, A.K. Dahle, D.H. StJohn, Grain refinement of magnesium, Magnesium Technology 2000 (2000) 211-218.
[37] G. Mann, J.R. Griffiths, C.H. Caceres, Hall-Petch parameters in tension and compression in cast Mg-2Zn alloys, Journal of Alloys and Compounds 378 (2004) 188-191.
[38] X. Gao, J.F. Nie, Characterization of strengthening precipitate phases in a Mg-Zn alloy, Scripta Materialia 56 (2007) 645-648.
[39] J.B. Clark, Transmission electron microscopy study of age hardening in a Mg-5 wt.% Zn alloy, Acta Metallurgica 13 (1965) 1281-1289.
[40] L. Wei, G. Dunlop, H. Westengen, Precipitation Hardening of Mg-Zn and Mg-Zn-RE alloys, Metallurgical and Materials Transactions A 26 (1995) 1705-1716.
[41] A.A. Nayeb-Hashemi, J.B. Clark, Phase diagrams of binary magnesium alloys, ASM International, Metals Park, Ohio, 1988.
[42] A. Akhtar, E. Teghtsoonian, Substitutional solution hardening of magnesium single crystals, Philosophical Magazine (1972) 897-916.
[43] E.D. Levine, W.F. Sheely, R.R. Nash, Solid-solution strengthening of magnesium single crystals at room temperature, Transactions of the Metallurgical Society of AIME 215 (1959) 521-526.
[44] A.H. Blake, C.H. Cáceres, Solid-solution hardening and softening in Mg-Zn alloys, Materials Science and Engineering: A 483-484 (2008) 161-163.

7. Bibliography

[45] Z. Hildebrand, M. Qian, D.H. StJohn, M.T. Frost, Influence of zinc on the soluble zirconium content in magnesium and the subsequent grain refinement by zirconium, Magnesium Technology 2004 (2004) 241-245.

[46] Y.C. Lee, A.K. Dahle, D.H. StJohn, The role of solute in grain refinement of magnesium, Metallurgical and Materials Transactions A 31A (2000) 2895-2906.

[47] M. Qian, D.H. StJohn, M.T. Frost, Effect of soluble and insoluble zirconium on the grain refinement of magnesium alloys, Materials Science Forum 419-422 (2003) 593-598.

[48] J. Bohlen, D. Letzig, K.U. Kainer, New perspectives for wrought magnesium alloys, Materials Science Forum 546-549 (2007) 1-10.

[49] J. Swiostek, J. Göken, D. Letzig, K.U. Kainer, Hydrostatic extrusion of commercial magnesium alloys at 100 °C and its influence on grain refinement and mechanical properties, Materials Science and Engineering: A 424 (2006) 223-229.

[50] T.B. Massalski, Binary Alloy Phase Diagrams, ASM International, 1992.

[51] G.I. Morozova, V.V. Tikhonova, N.F. Lashko, Phase composition and mechanical properties of cast Mg−Zn−Zr alloys, Metal Science and Heat Treatment 20 (1978) 657-660.

[52] S. Zhang, Phase constitution and morphologies of Mg-Zn-Zr alloy (MB15) Acta Metallurgica Sinica 3 (1990) 110-115.

[53] R. Arroyave, Z.K. Liu, Thermodynamics of Mg-Zn-Zr: implication on the effect of Zr on the grain refining of Mg-Zn alloys, Magnesium Technology 2005 (2005) 203-208.

[54] G. Petzow, G. Effenberg, Ternary Alloys, VCH Verlagsgesellschaft, Weinheim, Germany, 2000.

[55] L.L. Rokhlin, Magnesium alloys containing rare earth metals, Taylor & Francis, London, UK, 2003.

[56] C. Ma, M. Liu, G. Wu, W. Ding, Y. Zhu, Tensile properties of extruded ZK60-RE alloys, Materials Science and Engineering A 349 (2003) 207-212.

[57] Z.P. Luo, D.Y. Song, S.Q. Zhang, Strengthening effects of rare earths on wrought Mg-Zn-Zr-RE alloys, Journal of Alloys and Compounds 230 (1995) 109-114.

[58] G. Neite, K. Kubota, K. Higashi, F. Hehmann, Magnesium - Based Alloys, in: K.H. Matucha (Ed.), Structure and Properties of Nonferrous Alloys, Wiley - VCH, Weinheim, 1996, pp. 113 - 212.

[59] K.U. Kainer, J. Bohlen, D. Letzig, S. Mueller, K. Mueller, W. Reimers, Extruded magnesium alloys for structural applications, 64th Annual World Magnesium Conference Proceedings of the International Symposium on Magnesium Technology in the Global Age (2007).

[60] K.U. Kainer, J. Bohlen, D. Letzig, Status of the development of new wrought magnesium alloys for transportation, Materials Science Forum 539-553 (2006).

[61] K. Laue, H. Stenger, Extrusion: Processes, Machinery, Tooling, First ed., American Society for Metals 1981.

[62] S. Mueller, K. Mueller, H. Tao, W. Reimers, Microstructure and mechanical properties of the extruded Mg-alloys AZ31, AZ61, AZ80, International Journal of Materials Research 97 (2006) 1384-1391.

[63] J. Bohlen, F. Chmelik, P. Dobron, D. Letzig, P. Lukac, K.U. Kainer, Acoustic emission during tensile testing of magnesium AZ alloys, Journal of Alloys and Compounds 378 (2004) 214-219.

[64] K. Mueller, Direct and indirect extrusion of AZ31, TMS, Magnesium technology 2002 (2002).

[65] T. Murai, S.-i. Matsuoka, S. Miyamoto, Y. Oki, Effects of extrusion conditions on microstructure and mechanical properties of AZ31B magnesium alloy extrusions, Journal of Materials Processing Technology 141 (2003) 207-212.

7. Bibliography

[66] J. Bohlen, J. Swiostek, W.H. Sillekens, P.-J. Vet, D. Letzig, K.U. Kainer, Process and alloy development for hydrostatic extrusion of magnesium: The European Community research project MAGNEXTRUSCO Magnesium Technology 2005 (2005) 241-246.

[67] J. Swiostek, J. Bohlen, D. Letzig, K.U. Kainer, Hydrostatic and Indirect Extrusion of AZ-Magnesium Alloys, Materials Science Forum 488-489 (2005) 491-494.

[68] J. Bohlen, J. Swiostek, H.G. Brokmeier, D. Letzig, K.U. Kainer, Low temperature hydrostatic extrusion of magnesium alloys Magnesium Technology 2006 (2006) 213-217.

[69] M. Bauser, Metallkundliche Grundlagen, in: Strangpressen, Aluminium-Verlag, Duesseldorf, 2001.

[70] T. Murai, S.-i. Matsuoka, S. Miyamoto, Y. Oki, S. Nagao, H. Sano, Effects of zinc and manganese contents on extrudability of Mg-Al-Zn alloys, Journal of Japan Institute of Light Metals 53 (2003) 27-31.

[71] T.E. Leontis, Effect of rare-earth metals on the properties of extruded magnesium, Journal of metals (AIME) 3 (1951) 987-993.

[72] Y. Zhang, X. Zeng, L. Liu, C. Lu, H. Zhou, Q. Li, Y. Zhu, Effects of yttrium on microstructure and mechanical properties of hot-extruded Mg-Zn-Y-Zr alloys, Materials Science and Engineering A 373 (2004) 320-327.

[73] F.J. Humphreys, A unified theory of recovery, recrystallization and grain growth, based on the stability and growth of cellular microstructures--I. The basic model, Acta Materialia 45 (1997) 4231-4240.

[74] F.J. Humphreys, A unified theory of recovery, recrystallization and grain growth, based on the stability and growth of cellular microstructures--II. The effect of second-phase particles, Acta Materialia 45 (1997) 5031-5039.

[75] M.M. Myshlyaev, H.J. McQueen, A. Mwembela, E. Konopleva, Twinning, dynamic recovery and recrystallization in hot worked Mg-Al-Zn alloy, Materials Science and Engineering A 337 (2002) 121-133.

[76] G. Gottstein, Physical Foundations of Materials Sciences, Springer-Verlag, Berlin Heidelberg, Germany, 2004.

[77] L.W.F. Mackenzie, G.W. Lorimer, F.J. Humphreys, T. Wilks, Recrystallization behaviour of two magnesium alloys, Materials Science Forum 477-482 (2004) 477-482.

[78] L.W.F. Mackenzie, F.J. Humphreys, G.W. Lorimer, K. Savage, T. Wilks, Recrystallization behaviour of four magnesium alloys, Proceedings of the 6th International Conference Magnesium Alloys and Their Applications (2004) 158-163.

[79] M.T. Pérez-Prado, J.A. del Valle, O.A. Ruano, Effect of sheet thickness on the microstructural evolution of an Mg AZ61 alloy during large strain hot rolling, Scripta Materialia 50 (2004) 667-671.

[80] O. Sitdikov, R. Kaibyshev, T. Sakai, Dynamic recrystallization based on twinning in coarse-grained Mg, Materials Science Forum 419-422 (2003) 521-526.

[81] F.J. Humphreys, M. Hatherley, Recrystallization and related annealing phenomena Elsevier, United Kingdom, 2004.

[82] E.A. Calnan, C.J.B. Clews, The development of deformation textures in metals, Part III. Hexagonal structures, The Philosophical Magazine-A: Physics of Condensed Matter Structure Defects and Mechanical Properties 42 (1951) 919-931.

[83] A. Beck, H. Altwicker, Magnesium und seine Legierungen, 2nd. ed., Springer, 1939.

[84] I.L. Dillamore, W.T. Roberts, Preferred orientation in wrought and annealed metals, Metallurgical Reviews 10 (1965) 300-307.

[85] S. Yi, Investigation on the deformation behavior and the texture evolution in magnesium wrought alloy AZ31, Fakultät für Natur- und Materialwissenschaften, Technischen Universtität Clausthal, 2005.

7. Bibliography

[86] J. Bohlen, S.B. Yi, J. Swiostek, D. Letzig, H.G. Brokmeier, K.U. Kainer, Microstructure and texture development during hydrostatic extrusion of magnesium alloy AZ31, Scripta Materialia 53 (2005) 259-264.
[87] S. Müller, K. Müller, M. Rosumek, W. Reimers, Microstructure development of differently extruded Mg alloys, Part II ALUMINIUM 82 (2006) 438-442.
[88] J. Bohlen, J. Swiostek, D. Letzig, K.U. Kainer, Influence of the alloying additions on the microstructure development of extuded Mg-Mn alloys, Magnesium Technology 2009 (2009) 225-230.
[89] N. Stanford, M. Barnett, Effect of composition on the texture and deformation behaviour of wrought Mg alloys, Scripta Materialia 58 (2008) 179-182.
[90] N. Stanford, M.R. Barnett, The origin of "rare earth" texture development in extruded Mg-based alloys and its effect on tensile ductility, Materials Science and Engineering: A In Press, Corrected Proof (2008).
[91] S.R. Agnew, P. Mehrotra, T.M. Lillo, G.M. Stoica, P.K. Liaw, Crystallographic texture evolution of three wrought magnesium alloys during equal channel angular extrusion, Materials Science and Engineering: A 408 (2005) 72-78.
[92] S.B. Yi, H.G. Brokmeier, J. Bohlen, D. Letzig, K.U. Kainer, Neutron diffraction study on the texture development during extrusion of magnesium alloy AZ31, Physica B: Condensed Matter 350 (2004) E507-E509.
[93] M.T. Pérez-Prado, O.A. Ruano, Texture evolution during annealing of magnesium AZ31 alloy, Scripta Materialia 46 (2002) 149-155.
[94] S. Agnew, D.W. Brown, C.N. Tome, Validating a polycrystal model for the elastoplastic response of magnesium alloy AZ31 using in situ neutron diffraction, Acta Materialia 54 (2006) 4841-4852.
[95] D. Letzig, J. Swiostek, J. Bohlen, P.A. Beaven, Extrusion of AZ-series magnesium alloys, Proceedings of the International Symposium on Magnesium Technology in the Global Age (2006) 569-580.
[96] J. Bohlen, P. Dobron, E. Meza-García, F. Chmelík, P. Lukác, D. Letzig, K.U. Kainer, The effect of grain size on the deformation behaviour of magnesium alloys investigated by the acoustic emission technique, Advanced Engineering Materials 8 (2006) 422-427.
[97] M. Bauser, G. Sauer, K. Siegert, Extrusion, Second ed., ASM International, 2001.
[98] B. Avitzur, Handbook of metal-forming processes, John Wiley & Sons Inc New York, 1983.
[99] J. Swiostek, Erweiterung der Prozessgrenzen beim Strangpressen von Magnesiumknetlegierungen der AZ-Reihe durch das Hydrostatische Strangpressverfahren Technische Universität Hamburg-Harburg 2008.
[100] U.F. Kocks, C.N. Tomé, H.-R. Wenk, Texture and Anisotropy: Preferred Orientations in Polycrystals and their Effect on Materials Properties, Cambridge University Press, Cambridge, UK, 1998.
[101] V. Kree, J. Bohlen, D. Letzig, K.U. Kainer, Pract. Metallogr. 41 (2004) 233-246.
[102] DIN EN 50600, DIN-Materialprüfnormen für metallischen Werkstoffe 1 Metallographische Gefügebilder, Abbildungsmaßstäbe und Formate (1980-03).
[103] DIN EN 50125, DIN-Materialprüfnormen für metallischen Werkstoffe 1 (2000) 81-88.
[104] DIN EN 50106, DIN-Materialprüfnormen für metallischen Werkstoffe 1 (2000) 54-80.
[105] S.B. Yi, Private communications, 2008.
[106] Z. Luo, S. Zhang, Microstructures of Mg-Zr, Mg-Zn and Mg-Zn-Zr alloys, Acta Metallurgica Sinica 6 (1993) 337-342.

7. Bibliography

[107] W.-h. Wu, C.-q. Xia, Microstructures and mechanical properties of Mg-Ce-Zn-Zr wrought alloy, Journal of Central South University of Technology 11 (2004) 367-370.

[108] M.A. Easton, D.H. StJohn, A model of grain refinement incorporating alloy constitution and potency of heterogeneous nucleant particles, Acta Materialia 49 (2001) 1867-1878.

[109] A. Becerra, M. Pekguleryuz, Effects of zinc, lithium, and indium on the grain size of magnesium, Journal of Materials Research 24 (2009) 1722-1729.

[110] M.A. Easton, D.H. StJohn, Improved prediction of the grain size of aluminium alloys that includes the effect of cooling rate, Materials Science and Engineering: A 486 (2008) 8-13.

[111] Q. Ma, D.H. StJohn, M.T. Frost, Characteristic zirconium-rich coring structures in Mg-Zr alloys, Scripta Materialia 46 (2002) 649-654.

[112] C. Harris, Q. Li, M.R. Jolly, Prediction of extruded microstructures using experimental and numerical modelling techniques, Aluminium Two Thousand, 5th World Congress (2004).

[113] D.L. Atwell, Influence of alloying additions on the extrudability of magnesium, School of engineering and technology, Deaking University, 2005.

[114] H.L. Byoung, N.S. Reddy, J.T. Yeom, C.S. Lee, Flow softening behavior during high temperature deformation of AZ31 Mg alloy, Journal of Materials Processing Technology 187-188 (2007) 766-769.

[115] T. Laser, C. Hartig, R. Bormann, J. Bohlen, D. Letzig, Dynamic recrystallization of Mg-3Al-1Zn, Proceedings of the 6th International Conference Magnesium Alloys and Their Applications (2004) 164-169.

[116] I. Stloukal, J. Cermak, Diffusion of Zn in two-phase Mg-Al alloy, Defect and Diffusion Forum 263 (2007) 189-194.

[117] I.L. Dillamore, P. Hadden, D.J. Stratford, Texture control and the yield anisotropy of plane strain magnesium extrusions, Texture 1 (1972) 17-29.

[118] E. Meza-García, J. Bohlen, D. Letzig, K.U. Kainer, Influence of alloying elements on the mechanical properties of zinc containing extruded magnesium alloys Magnesium Technology 2007 (2007) 263-268.

I want morebooks!

Buy your books fast and straightforward online - at one of world's fastest growing online book stores! Environmentally sound due to Print-on-Demand technologies.

Buy your books online at
www.morebooks.shop

Kaufen Sie Ihre Bücher schnell und unkompliziert online – auf einer der am schnellsten wachsenden Buchhandelsplattformen weltweit! Dank Print-On-Demand umwelt- und ressourcenschonend produziert.

Bücher schneller online kaufen
www.morebooks.shop

KS OmniScriptum Publishing
Brivibas gatve 197
LV-1039 Riga, Latvia
Telefax: +371 686 204 55

info@omniscriptum.com
www.omniscriptum.com

Printed by Books on Demand GmbH, Norderstedt / Germany